BAYESIAN NETWORKS - ADVANCES AND NOVEL APPLICATIONS

Edited by **Douglas McNair**

Bayesian Networks - Advances and Novel Applications
http://dx.doi.org/10.5772/intechopen.75254
Edited by Douglas McNair

Contributors

Rosa Arnaldo, Alvaro Rodriguez Sanz, Victor Fernando Gomez Comendador, Eduardo Sanchez Ayra, Javier Alberto Perez Castan, Luis Perez Sanz, Bouchra Zoullouti, Mustapha Amghar, Nawal Sbiti, Mirko Perkusich, Luiz Antonio Silva, João Nunes, Kyller Gorgônio, Hyggo Almeida, Angelo Perkusich, Cristiano Premebida, Diego Faria, Luis Manso, Eduardo Parente Ribeiro, Pedro Núñez, Oleg Kupervasser, Septia Devi Prihastuti Yasmirullah, Nur Iriawan, David Olayungbo, Douglas McNair

Notice

First published in London, United Kingdom, 2019 by IntechOpen
IntechOpen is the global imprint of INTECHOPEN LIMITED, registered in England and Wales, registration number: 11086078, 7th floor, 10 Lower Thames Street, London,
EC3R 6AF, United Kingdom
Printed in Croatia

British Library Cataloguing-in-Publication Data
A catalogue record for this book is available from the British Library

Additional hard and PDF copies can be obtained from orders@intechopen.com

Bayesian Networks - Advances and Novel Applications, Edited by Douglas McNair
p. cm.
Print ISBN 978-1-83962-322-6
Online ISBN 978-1-83962-323-3
eBook (PDF) ISBN 978-1-83962-324-0

For EU product safety concerns:
IN TECH d.o.o., Prolaz Marije Krucifikse Kozulić 3, 51000 Rijeka, Croatia,
info@intechopen.com or visit our website at intechopen.com.

Meet the editor

Douglas McNair serves as a senior advisor in quantitative sciences—analytics innovation in global health at the Bill & Melinda Gates Foundation. He assists the foundation's R&D in drug and vaccine development for infectious diseases, childhood diseases, and neglected tropical diseases. Current projects include product development programs in discovery and translational sciences involving Bayesian networks. His activity also includes machine-learning and modeling of health economics collaborating with the Global Development Division. Previously, Douglas was President of Cerner Math Inc., responsible for artificial intelligence (AI) components of Cerner's EHR solutions, discovering AI predictive models from real-world de-identified EHR-derived Big Data. Douglas is lead inventor on more than 100 patents and pending patent applications, including several involving Bayesian predictive models for clinical diagnostics.

Contents

Preface

In recent years, Bayesian networks have experienced increased interest and widely varied applications in numerous areas, including economics, risk analysis and assets and liabilities management, AI and robotics, transportation systems planning and optimization, political science analytics, law and forensic science assessment of agency and culpability, pharmacology and pharmacogenomics, systems biology and metabolomics, psychology, and policy-making and social programs evaluation. This strong and diverse response results not least from the fact that plausibilistic Bayesian models of structures and processes can be robust and stable representations of causal relationships. Such stability and resilience to multi-sourced data has allowed to design practical solutions that yield important and novel insights. Additionally, Bayesian networks' amenability to incremental or longitudinal improvement through incorporating new data affords extra advantages compared to traditional frequentist statistical methods. We have created this volume with a view toward colleagues in the field of machine learning and Bayesian networks and to students at the graduate or postgraduate level.

In terms of epistemology, Bayesian networks promise to help achieve improved accuracy regarding the truth of propositions of interest and regarding the causal and statistical basis for their truth. Moreover, Bayesian networks can reveal relationships that have face-validity to decision-makers and the public. To the degree that they illuminate a credible basis for particular probabilistic solutions, such improvements can enable setting forth mechanisms and principles in a defensible way, supported by a basis that can anchor just and stable policy. The present volume includes new contributions from a number of innovators in Bayesian networks with emphasis on socially important applications.

In the Introduction, I call attention to the contemporary relevance of Bayesian networks, in a time and culture that needs epistemological 'ground truth', or as near as one can come to this, as a basis for rational, ethical management of social, engineering, and biological systems. Bayesian networks are particularly effective in this connection insofar as the arcs that are empirically learned or induced for the networks very often accurately represent the direction of causality. Events or states that share a de facto cause are likely to be conditionally independent given the cause; arrows in the causal direction capture this independence. In a naïve Bayes network, the arcs are often not in the right causal direction (e.g., diabetes does not cause aging). But in non-naïve and other types, the arcs are highly accurate regarding causality. This aspect is not only valuable as to low error rates in practical applications but also affords a greater degree of face-validity, transparency, and social and psychological acceptability compared to certain other machine-learning methods and AI model types.

Chapter 2 by Septia Yasmirullah and Nur Iriawan concerns growth model approaches using hierarchical Bayesian methods. They address emerging economic imbalances within Indonesian regions, subsequent to 2004. Economic development naturally entails issues of fairness and equitability of resources application as well as determinations of programs' efficacy. Accurate analyses to identify disparities and their causes are therefore fundamental to justice and value creation in terms of public policy-making and program evaluation. A valuable finding useful to others in the field is that one-level Bayesian modeling, with regional attributes ignored and only national attributes retained, while Bayesian hierarchical structure modeling, accounting as it does for regional attributes, are able to accurately characterize interactions through the modeling of parameters of micro models with province-level covariates, achieving smaller DIC values and better explanatory power. The authors' finding that inter-regional variations are significantly affected by the regions' cities and province characteristics confirms natural intuitions and serves to further recommend the advantages and strengths of the more comprehensive hierarchical method. The authors' work ably serves the important public-sector goal in most countries, to enhance social wellbeing through effective economic development in a manner that is trans-regionally fair and just.

Chapter 3 by Rosa Maria Arnaldo, Victor Fernando Gómez Comendador, Alvaro Rodriguez Sanz, Eduardo Sanchez Ayra, Javier Alberto Pérez Castán, and Luis Perez Sanz presents an approach for accurate analysis and engineering of aviation under conditions of uncertainty. Commencing with a cogent survey of the work in applying Bayesian networks to aviation and transportation industry decision-making, the authors show that Bayesian methods can be used to select or parameterize input distributions for a probabilistic model. In this case, the decision process may be made more acceptable in public policy or regulatory regimes that are historically guided by frequentist statistics or are not yet reconciled to modern Bayesian methods.

Chapter 4 by Bouchra Zoullouti, Nawal Sbiti, and Mustapha Amghar examines Bayesian network analysis for descriptive and quantitative characterizations of risks associated with health services. This approach allows to construct risk matrices concerning the safety outcomes of interest and accurately predict patients' likelihood of experiencing each adverse event or outcome. As such, Bayesian networks can be a means not only of guiding structure and process improvements but also of individualizing therapeutic strategies. This promises to become a valuable modality in contemporary precision medicine initiatives.

In recent work on operationalizing Bayesian networks, it is recognized that it is impossible to achieve a reliably accurate description of the processes involved without new data being collected or a large amount of data being stored that cannot be analyzed at once. Chapter 5 by Mirko Perkusich summarizes and elaborates some of the findings on the design of continuous learning Bayesian networks to accommodate and adapt to new incoming information. Statistical power sufficient to justify model updating is an important issue that has not been adequately covered in the research literature to date. To convert batch processing to a continuous process for model updating one additionally needs to reconcile the rate of data accrual with the perhaps short amount of time for updating the Bayesian network, a case where the algorithm only updates the posterior probabilities of the parent lattices. Friedman-Goldszmidt, Lam-Bacchus, Roure, and Shi-Tan serial Bayesian network updating are examined, and a hybrid algorithm that offers computational complexity and accuracy advantages is proposed.

Chapter 6 by Pedro Núñez, Eduardo Parente Ribeiro, Luis Manso, and Cristiano Premebida analyzes the impact of fusion of disparate datatypes and sources (video and still cameras, LiDAR, wearable sensors, trading transaction timeseries, etc.) on emulation of competent human perception. This is important in regard to mobile robotics, advanced driver assistance systems, multimodal sensor fusion for object detection, and a wide variety of other AI and robotics applications. In this era of autonomous driving systems and other technologies on which is conferred life-critical autonomy, the notion that perception involves Bayesian inference is an increasingly popular position, one that merits elaboration and detailed critical, quantitative study.

Chapter 7 by Oleg Kupervasser provides a detailed account of a Bayesian network basis for evaluating medicinal chemistry (quantitative structure-activity relationships, QSAR, physiologically based pharmacokinetics, etc.) in drug discovery and development. The emerging importance of Bayesian network methods derives partly from the difficulty and inaccuracies of present quantum chemical models and from the impracticality of sufficient characterization of structure of drug molecules and receptor active sites, including vicinal waters in and around hydrophobic pockets in active sites. This is particularly so for biologicals (protein and nucleic acid APIs) and target applications that exhibit extensive inter-receptor trafficking, genomic polymorphisms, and other system biology phenomena. The effectiveness and accuracy of Bayesian methods for drug development likewise depends on certain prerequisites, such as an adequate distance metric by which to measure similarity/difference between combinatorial library molecules and known successful ligand molecules targeting a particular receptor and addressing a particular clinical indication. In this connection, the distance metric proposed in the chapter and the associated Lemmas and Proofs are of substantial value in the future of high throughput screening (HTS), combinatorial library management, and medicinal chemistry in the era of precision medicine and personalized genomics-informed pharmaceutics. Today there is a growing number of compounds and clinical indications for which de-risked "hit-to-lead" and "lead-to-candidate" prediction and decision-making have been successfully accomplished using Bayesian network methods and distance metrics. In this regard, the methods have growing interest for assisting in drug development and foundations', NGOs', and private venture financing of drug discovery and M&A, particularly in indications that involve multi-receptor cross-talk, metabolomic cascades, rare and neglected diseases, and other complex system biology.

Bayesian network modeling in macroeconomics, illustrated by a 37-year epoch in the recent experience of Nigeria, is the focus of the concluding chapter by David Olayungbo. This contribution elucidates what is, to date, an under-appreciated application of Bayesian methods in a manner than can help guide economic policy-setting and fiscal program evaluation at the societal level. Bayesian network models for competitiveness growth in economies reveal and reflect the dynamics and the results of antecedent policies – the constituents of the conditions and the context for subsequent growth – and thereby inform stakeholders and decision-makers regarding [causal] relationships bearing on alternative means for achieving competitive catch-up or continued success.

I would like to thank and express gratitude to all the authors for their contributions. I wish readers a fruitful and enlightening reading.

Douglas S. McNair MD PhD
Senior Advisor
Quantitative Sciences - AI & Knowledge Integration
Bill & Melinda Gates Foundation
Seattle, Washington, USA

Chapter 1

Introductory Chapter: Timeliness of Advantages of Bayesian Networks

Douglas S. McNair

Additional information is available at the end of the chapter

http://dx.doi.org/10.5772/intechopen.83607

1. The timeliness of Bayesian networks in an era of problematized truth-claims

As a child, I was raised as a Lutheran, with an earnest interest and concern for scripture. I became notorious for asking my Sunday school teachers imponderable and impolitic questions. Upon encountering Genesis 3:11–13 around age 6, I noticed that God confronts Adam in the Garden of Eden and asks, "Have you eaten from the tree?" Adam prevaricates: "The woman whom you gave to be with me, she gave me fruit from the tree." God inquires of Eve about this. She answers, "The serpent tricked me." My youngster mind recognized this pattern of dialog as very much akin to my own defensive dissembling with my parents when I had been the cause of some accident or had done something wrong. I very much wanted to know why Adam's and Eve's reasoning was insufficient.

God asks "what," but humans typically answer with proposals as to "why" (see [1], p. 24). We humans crave reasons and often we value causal explanations far more than facts. Adam and Eve believed that identifying causes outside themselves would exculpate them. This notion can be wrong, but the tendency in our species is strong and is an important aspect of computational ("artificial") intelligence, particularly in the present era in which AI is being widely deployed and operationalized and bestowed with progressively greater autonomy and influence over our lives. The present volume is motivated in part in recognition of this trend and the fact that social acceptance of AI strongly depends upon transparency and face-valid explanations that justify or satisfactorily legitimate authority that is exerted over us.

Bayesian networks (BNs) have come a long way since Rev. Bayes' original paper [2], and the applications in which they excel are by now very diverse. For example, Kass and Raftery [3] set forth a summation of dozens of uses for, interpretations of, and advantages and disadvantages of Bayes factors in hypothesis testing. Bayesian networks represent graphically uncertainties and decisions that expressly represent the relationships and the strengths of probabilistic dependences among the variables and the associated information flows. A chief advantage of BNs is that they allow to address uncertainties and evidence from disparate sources, such as expert judgment [4–6] and observable experience, being able to take into account common causes and influences of social and logistical aspects [7]. In BNs, variables and their interdependencies are encoded as nodes and directed arcs with conditional probability tables (CPTs) linked with the nodes. Under the assumption of conditional independence, a BN represents the joint probability distribution of variables [8]. Bayesian networks, besides this natural transparency, revealing joint dependencies graphically in directed acyclic graphs (DAGs) whose nodes denote elements or factors associated with concepts that we can name and understand, have the notable advantage of modeling causality (more conveniently than by other methods, e.g., transfer entropy, Granger asymmetric/noncommutative correlation, etc.), in a manner that yields empirically credible transmission of evidence or influence. This capability in turn produces stochastic classifiers that can be combined with utility functions to automate optimal decision-making that emulates the decision-making embodied in the data from which the BN was learned.

2. Bayesian networks in use-cases involving epistemological or perceptual complexity

Other attributes of BNs that are timely and valuable for contemporary use-cases and applications include:

- facilitate incorporating causal knowledge resulting in probabilities that are easy to explain;
- enable consistent combining of information from various sources (including expert elicitation and crowd-sourcing) and mixed data types;
- batch or continuous updating that can be responsive to newly acquired or incoming data;
- amenable to processes aimed at measuring and accounting for model structural uncertainty;
- amenable to modeling partially observed and unlabeled data; and
- can estimate certainties for the values of variables that are not observable (or whose cost or rate of change limits the extent or frequency of direct observation).

Bayesian networks function most effectively when the arcs that are learned or induced for the BN accurately represent the direction of causality. Events or states that share a common cause are likely to be conditionally independent given the cause; arrows in the causal direction capture this independence. Adam's and Eve's (and our) human nature and mortal susceptibility to temptation were (are) common causes in just such a way, in a manner that even a child could grasp [9–11]. In a naïve Bayes network, the arcs are often not in the right causal direction (e.g., diabetes does not cause aging). But in non-naïve and other BN types, the arcs are mostly accurate regarding causality (e.g., diabetes does cause insulin to be low or insulin sensitivity to be low), and this feature is sufficient to make such BNs not only useful but humanly understandable and socially endorsable, even in highly complex contexts [12–23].

Contributions in other chapters in the present volume explore a variety of novel ways in which BNs are becoming ever more relevant and impactful within the broader armamentarium of AI methods for real-world applications. My own recent engagement with Bayesian networks has been primarily directed to pharmacogenomics-related systems biology and physiologically based pharmacokinetics (PBPK) modeling for efficient drug development and personalized medicine. However, the aspect of credible (Bayesian) accounts of causation that were timely and salient to me at age 6 remain so now 60 years later and are exemplified by contemporary BNs. I anticipate that readers will likely find them so as well.

"A model is a simplification or approximation of reality and hence will not reflect all of reality. ... [George E. P. Box] noted that 'All models are wrong, but some are useful.' While a model can never be [full, immutable, ground] 'Truth,' a model might be ranked from very useful, to useful, to somewhat useful to, finally, essentially useless." —Kenneth Burnham and David Anderson (2002).

Author details

Douglas S. McNair

Address all correspondence to: douglas.mcnair@gatesfoundation.org

Quantitative Sciences - AI & Knowledge Integration, Bill & Melinda Gates Foundation, Seattle, Washington, USA

References

[1] Pearl J, Mackenzie D. The Book of Why. New York: Basic Books; 2018

[2] Bayes T. An essay toward solving a problem in the doctrine of chances. Philosophical Transactions of the Royal Society of London. 1763;**53**:370-418

[3] Kass RE, Raftery AE. Bayes factors. Journal of the American Statistical Association. 1995;**90**:773-795

[4] Albert I, Donnet S, Guihenneuc-Jouyaux C, Low-Choy S, Mengersen K, Rousseau J. Combining expert opinions in prior elicitation. Bayesian Analysis. 2012;**7**:503-532

[5] Morris DE, Oakley JE, Crowe JA. A web-based tool for eliciting probability distributions from experts. Environmental Modelling & Software. 2014;**52**:1-4

[6] Pitchforth J, Mengersen K. A proposed validation framework for expert elicited Bayesian networks. Expert Systems with Applications. 2013;**40**:162-167

[7] Ale B, Van Gulijk C, Hanea A, Hanea D, Hudson P, Lin P-H. Towards BBN based risk modelling of process plants. Safety Science. 2014;**69**:48-56

[8] Jensen FV, Nielsen TD. Bayesian Networks and Decision Graphs. 2nd ed. New York: Springer; 2007

[9] Andrews M, Baguley M. Prior approval: The growth of Bayesian methods in psychology. British Journal of Mathematical and Statistical Psychology. 2013;**66**:1-7

[10] Burnham KP, Anderson DR. Model Selection and Multimodel Inference: A Practical Information-Theoretic Approach. 2nd ed. New York: Springer-Verlag; 2011

[11] Fenton N, Neil M. Risk Assessment and Decision Analysis with Bayesian Networks. 2nd ed. London: Chapman & Hall; 2018

[12] Grover J. The Manual of Strategic Economic Decision Making: Using Bayesian Belief Networks to Solve Complex Problems. New York: Springer; 2016

[13] Kelly D, Smith C. Bayesian Inference for Probabilistic Risk Assessment. London: Springer; 2011

[14] Kjærulff UB, Madsen AL. Bayesian Networks and Influence Diagrams: A Guide to Construction and Analysis. New York: Springer; 2008

[15] Korb K, Nicholson AE. Bayesian Artificial Intelligence. 2nd ed. London: Chapman & Hall; 2010

[16] Maathuis M, Drton M, Lauritzen S, Wainwright M, editors. Handbook of Graphical Models. London: Chapman & Hall; 2018

[17] Morgan MG. Use (and abuse) of expert elicitation in support of decision making for public policy. Proceedings of the National Academy of Sciences of the United States of America. 2014;**111**:7176-7184

[18] Neapolitan RE. Probabilistic methods for bioinformatics: With and introduction to Bayesian networks. Burlington, MA: Morgan Kaufmann; 2009

[19] Peace KE, Chen D-G, Menon S, editors. Biopharmaceutical Applied Statistics Symposium. New York: Springer; 2018

[20] Causality PJ. Models, Reasoning, and Influence. 2nd ed. Cambridge, UK: Cambridge University Press; 2009

[21] Raftery AE. Approximate Bayes factors for accounting for model uncertainty in generalized linear models. Biometrika. 1996;**183**:251-266

[22] Suzuki J, Ueno M, editors. Advanced methodologies for Bayesian networks: Second International Workshop (AMBN 2015); Yokohama, Japan. New York: Springer; 2015

[23] Wilkinson D. Stochastic Modelling for Systems Biology. 3rd ed. London: Chapman & Hall; 2018

Chapter 2

An Economic Growth Model Using Hierarchical Bayesian Method

Nur Iriawan and Septia Devi Prihastuti Yasmirullah

Additional information is available at the end of the chapter

http://dx.doi.org/10.5772/intechopen.88650

Abstract

Economic growth can be used as an assessment for the success of the regional economic establishment. Since the Regulation of the Republic Indonesia Number 32 of 2004 has been implemented, the imbalance economic growth among the regencies in Indonesia is rising. The imbalance in the conditions of economic growth differs between regions with the aim of the government to improve social welfare by expanding economic activities in each region. The purpose of this chapter is to elaborate whether there is a difference in economic growth based on the distribution of bank credit for each regency in Indonesia. This research analyzes the economic growth data using hierarchical structure model that follows the normality-based modeling in the first level. The two modeling approaches will be applied, i.e., a general one-level Bayesian approach and a two-level structure hierarchical Bayesian approach. The success of these approaches has demonstrated that the two-level hierarchical structure Bayesian has a better estimation than a general one-level Bayesian. It demonstrates that all of the macro-level characteristics of provinces are significantly influencing the different economic growth in every related province. These variations are also significantly influenced by their cross-level interaction regency and provincial characteristics.

Keywords: Bayesian, estimation, economic growth, normal distribution, hierarchical

1. Introduction

The rising economic development in a country or in such region can be shown by its economic growth. It could be affected by three main factors, i.e., advances in technology, the capital accumulation of investment, and the local workforce participation [1]. The indicator to measure the economic growth rate and to determine the shifts and economic structural changes are the gross domestic product (GDP). There were two kinds of GDP, i.e., GDP at constant prices and GDP at current prices. GDP at constant prices was used to explain the economic growth from year to year, while GDP at current prices was used to see the economic structural changes [2].

The law of the Republic Indonesia Number 32 of 2004 states about the delegation of partial of the central government authority to the local government for conducting and organizing its own internal affairs. The increase of the economic activity in each regency and province in order to improve the national economy is the main goal of the delegation. Local autonomy welfare of society expected quickly can be realized through applied decentralization regulation. Decentralization, on the other hand, can drive the imbalances in economic growth among the regencies.

The Indonesian government issued the nine packages policy called Nawacita in 2014, a proposed solution to overcome the imbalance of economic growth. The nine packages policy in Nawacita consists of returning the state to have the main task of protecting all citizens and providing a safe living environment; emerging clean, effective, trusted, and good democratic governance; development of marginal areas; reforming law enforcement bureaus; improving life quality; increasing productivity and competitiveness; promoting economic independence by developing domestic strategic sectors; overhauling the nation character; and strengthening the spirit of unity in diversity and social reform [3].

The seventh of the nine points in Nawacita states that the government would accomplish economic independence by developing domestic strategic sectors. The economy sectors were stressed as a priority sector for accompanying Nawacita, which fitted the classification Indonesia Banking Statistics (named as Statistik Perbankan Indonesia or SPI). Economic growth is significantly affected by these sectors. A significant example is that the distribution of financial credit to economic priority sectors has been proven to have a significant contribution as a positive impact on regional economic growth [4].

As a developing country, the banking sector in Indonesia is still dominating the financial system. The development of the banking sector has a strong relationship with economic growth. Some previous studies have shown that there is a positive relationship between the number of bank credit with income per capita growth in both developed and developing countries [5, 6]. The banking industry characteristic in Indonesia, however, is believed relatively brittle [7], inefficient in financing intermediation in ASEAN [4].

This chapter discusses the bank credit influence on economic growth through an assessment of the distribution of financial credit in Indonesia using two-level Bayesian hierarchical structure modeling, each regency on the first level as a sample unit and provinces as the second level. There are 284 regencies as the selected sample unit from the first level, which spread unbalanced in the 11 selected provinces. Demonstration of the ability to resolve the challenges of modeling on this unbalance of a number of sample units, therefore, was a significant contribution of Bayesian hierarchical modeling. Different from the frequentist approaches, Bayesian analysis treats all unknown parameters as random variables which have distribution [8]. The results of this study are expected to provide guidance about financial credit distribution to priority sector and recommendation to policy-making in Bank Indonesia, the local government, Statistics Indonesia (BPS), and other related institutions.

2. Background and methodology

Economic growth has always been a benchmark for the success of the economic development of a country or a region. In a region, it can be conventionally measured by the increasing rate of the gross regional domestic product (GRDP) value represented in percent. The important indicators that represent the economic condition in a region for a certain reporting period were GRDP. There were two types of GRDP, i.e., reported as current prices and reported as constant prices. The performance of the economy over time in real terms could be seen through the GRDP at constant prices, while GRDP at current prices was used to see the shifts and the economic structures [2]. The economic growth rate in Indonesia during 2015 is lower than in 2014, i.e., 4.79% of 5.02%. Some provinces, however, have economic growth above national economic growth, which are West Sumatra, North Sumatra, East Java, Central Java, West Java, East Nusa Tenggara, Southeast Sulawesi, South Sulawesi, and Papua, while the economic growth rate of South Sumatra (4.5%) is lower than the national economic growth. In 2015, Papua Province was an exceptional province having the most fantastic rapid economic growth rate, amounting to 7.97%. The second most rapid economic growth rate after Papua Province belongs to South Sulawesi Province, achieving 7%. Five provinces were only able to reach the economic growth rate of around 5–6%, and one province exactly had a growth of 6%. The others grow below 5%. Almost all provinces that have GRDP at constant prices tend to increase from 2014 to the year 2015, except Aceh Province. The deficit balance of trade, foreign export oil, and imports are the main cause of decreasing its GRDP. They have reduced the level of Aceh's domestic economy (inter-regional) and sharpened differences in economic growth.

The secondary data recorded officially from the Economic Assessment and Surveillance Division of Economic and Financial Advisory, Bank Indonesia Representative Office of East Java Province, coupled with the data from Statistics Indonesia (BPS) are used in this study. There are 17 micro predictor variables (x), four macro predictor variables (w), and a response variable (y), i.e., economic growth rate. **Figure 1** shows the design of the hierarchical data structure. Due to the demand for the number of sample units as many as 17 variables in the modeling, only 11 provinces were used from as many as 34 provinces in Indonesia which had at least 16 regencies. This considers the guarantee in approaching the fulfillment of the requirements of the micro model that uses 17 predictor variables.

The procedures of analysis in this research follow the steps below:

1. Describing and exploring economic growth data each regency
2. Parameter estimation of a global one-level Bayesian model of all regencies
   a. Write an algorithm to estimate parameters of the general one-level Bayesian model.

      The parameters used for modeling in this study are τ and β. These are the parameters of the normal distribution. The preliminary procedure that needs to be done in

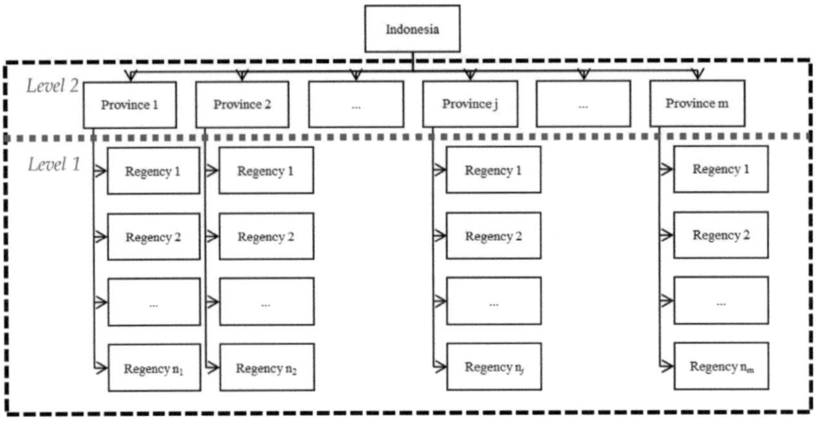

Figure 1. Hierarchical structure scheme.

modeling with the one-level Bayesian approach is to determine the prior distribution for the parameters to be estimated. This study uses an independent prior distribution, i.e., the prior distribution of each parameter is independent of one another. These independent prior distributions can be used to tackle problems in the modeling if it suspects there is high collinearity between the explanatory variables.

Prior distributions are used for each element of the parameter vector in the one-level Bayesian model-based normal distribution as follows:

$$
\begin{aligned}
&y_i = \beta_0 + \beta_1 x_{1i} + \beta_2 x_{2i} + \ldots + \beta_p x_{pi} + e_i,\\
&\mathbf{y} \sim N(\boldsymbol{\mu}, \sigma^2 \mathrm{I}),\\
&\boldsymbol{\mu} = \mathbf{x}\boldsymbol{\beta} + \mathbf{e},\\
&\tau = \frac{1}{\sigma^2},\\
&e_i \sim N(0, \sigma^2),\\
&\beta_s \sim N(\hat{\beta}_s, \sigma_s^2),\\
&\sigma_s^2 \sim Gamma(a_s, b_s),\\
&\text{where } i = 1, 2, \ldots, n; n \text{ is number of data,}\\
&\qquad s = 0, 1, 2, \ldots, p; p \text{ is number of micro predictor variables.}
\end{aligned}
\tag{1}
$$

Determining the value of hyper-parameter of each parameter in the prior distribution is done by a combination of the conjugate and pseudo priors [9]. This is done to ensure that the iteration of the parameter estimation process will quickly meet convergence and meet the properties of the Markov chain, i.e., irreducible, aperiodic, and recurrent.

b. Implement the algorithm into the syntax of WinBUGS and run.

The relationship between data and the prior distribution of parameters in Bayesian modeling can be illustrated as a graphic model form using a directed acyclic graph (DAG).

Figure 2 is a representation of the relationship among data, model parameters, and their parameter prior being modeled. Box-shaped node is used for representing the parameter or data which are constant, while the node ellipse is used for representing the parameters changing stochastically or as a logical structure relationship. Between nodes is connected by a single line and a dotted line. The single line is stating a stochastic relationship, while the dotted lines express logical relationships.

c. Analyze the model by listing significant contributions of each predictor variable using the concept of whether the zero value is inside the credible interval of its highest posterior distribution (HPD).

d. Measure the accuracy of this general one-level Bayesian model by computing its deviance information criterion (DIC) value.

3. Parameter estimation of the two-level hierarchical structure Bayesian model. The first-level model is for the regency level modeling, and the second-level model is for the province level modeling.

a. Write an algorithm to estimate parameters of the two-level hierarchical structure Bayesian model.

The hierarchical model parameter has a multilevel structure, called hyper-parameter. It is in line with the hierarchical design perspective in this problem, i.e., the hierarchy between regency and province. There are two parameters on the first level, namely, $\boldsymbol{\beta}$ and σ_y^2, and there are two parameters on the second level, i.e., γ and σ_{sj}^2. For the parameters in the first level, σ_y^2 represents the variance of normal error distribution, and $\boldsymbol{\beta}$ represents the parameters of regression in the micro model, while the parameters in the second level are referred to a hyper-parameter which is a prior distribution of the parameter $\boldsymbol{\beta}$. This parameter $\boldsymbol{\beta}$ will be set as a response in the regression model which is explained by hyper-parameter as a combination of the covariate w in the macro model.

The following important steps are determining the distribution and hyper-parameter prior for all of the parameters to be estimated. As in the global one-level model, in this two-level modeling, the independent prior distributions are used. Prior distributions

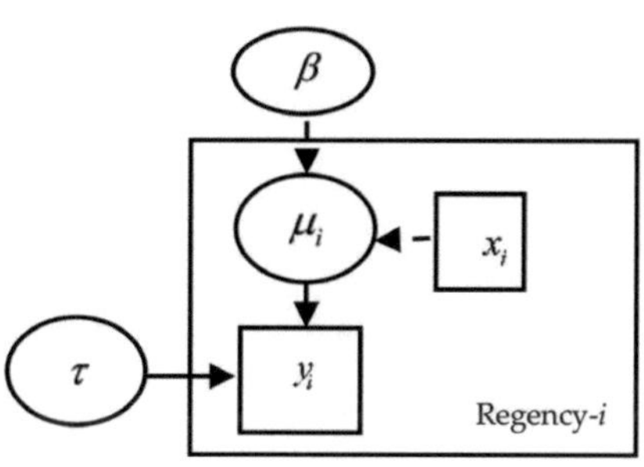

Figure 2. DAG one level Bayesian methods.

are used for each element of a Bayesian hierarchical model parameter vector based on the normal distribution as follows:

$$\begin{aligned}
&y_{ij} = \beta_{0j} + \beta_{1j}x_{1ij} + \beta_{2j}x_{2ij} + \ldots + \beta_{pj}x_{pij} + e_{ij},\\
&\mathbf{Y} \sim N\left(\boldsymbol{\mu}_{\mathbf{y}}, \sigma_y^2\mathbf{I}\right),\\
&\boldsymbol{\mu}_{\mathbf{y}} = \mathbf{x}\boldsymbol{\beta} + \mathbf{e},\\
&\beta_{sj} \sim N\left(\hat{\beta}_{sj}, \sigma^2_{\beta_{sj}}\right),\\
&e_{ij} \sim N\left(0, \sigma_y^2\right),\\
&\hat{\beta}_{sj} = \gamma_{0s} + \gamma_{1s}w_{1j} + \gamma_{2s}w_{2j} + \ldots + \gamma_{qs}w_{qj} + u_{sj},\\
&\hat{\boldsymbol{\beta}}_j \sim N\left(\boldsymbol{\mu}_{\beta j}, \sigma^2_{\beta j}\mathbf{I}\right),
\end{aligned} \tag{2}$$

$$\begin{aligned}
&\boldsymbol{\mu}_{\beta j} = \boldsymbol{\gamma}\mathbf{w} + \mathbf{u},\\
&\gamma_{ts} \sim N\left(\mu_{\gamma_{ts}}, \sigma^2_{\gamma_{ts}}\right),\\
&u_{sj} \sim N\left(0, \sigma^2_{\beta_{sj}}\right),\\
&\sigma^2_{\gamma_{ts}} \sim Gamma\left(a_{\gamma_{ts}}, b_{\gamma_{ts}}\right),\\
&\text{where } i = 1, 2, \ldots, n_j; j = 1, 2, \ldots, m;\\
&\qquad s = 0, 1, 2, \ldots, p; \text{ and } t = 0, 1, 2, \ldots, q.
\end{aligned}$$

As in the global one-level model, in this two-level modeling, the determining of the value of each parameter prior distribution is done by a combination of the conjugate and pseudo priors.

b. Implement the algorithm into the syntax of WinBUGS and run.

The hierarchical relationship of model parameters, i.e., parameter priors and hyper-parameter prior, in the Bayesian approach of such hierarchical scheme could be described by the directed acyclic graph [10, 11]. Data, parameters, and parameter prior models in the DAG are represented by nodes.

Figure 3 describes a Bayesian hierarchical model DAG for two-level model based on the normal distribution, i.e., the first level is the regency, and the second level is the provinces. For simplicity of writing, Regency-*i*, $i = 1, 2, \ldots, n_j$, where n_j is a number of regency in the *j*-th province, and Province-*j*, $j = 1, 2, \ldots, m$, where m is a number of the province. The parameter of regression in the first level is $\boldsymbol{\beta}$; it can be written individually as β_{sj}, where $s = 0, 1, 2, \ldots, p$; p is a number of the covariate in the micro model. While the parameters of regression in the second level is $\boldsymbol{\gamma}$, it can be written individually as γ_{ts}, where $t = 0, 1, 2, \ldots, q$; q is a number of the covariate in the macro model.

c. Analyze the first and second level model by creating a list of the significant contribution of predictor variables in each regency and province by using the concept of whether the zero value is inside the credible interval of its HPD.

http://dx.doi.org/10.5772/intechopen.88650

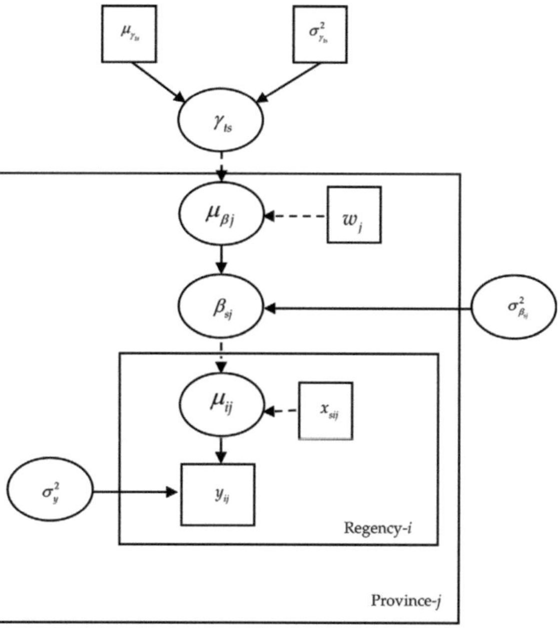

Figure 3. DAG hierarchical Bayesian methods.

d. Determine the accuracy of this two-level hierarchical structure Bayesian models by computing its DIC value.

4. Choosing the best model between the general one-level Bayesian model and two-level hierarchical structure Bayesian model by comparing their DIC values

The selection of the best models from the two models can use a smaller DIC value. DIC of the kth model can be determined through the following Equation [11]:

$$\begin{aligned} DIC(k) &= 2\overline{D(\theta_k, k)} - D(\overline{\theta}_k, k) \\ &= D(\overline{\theta}_k, k) + 2p_k \end{aligned} \tag{3}$$

where $D(\theta_m, m)$ $D(\theta_k, k)$ is a deviance that is equal to the negative value of twice the log-likelihood as stated in Eq. (4):

$$D(\theta_k, k) = -2\log f(y|\theta_k, k) \tag{4}$$

where $\overline{D(\theta_k, k)}$ is the average posterior and p_k represents the number of parameters in the kth model calculated as

$$p_k = \overline{D(\theta_k, k)} - D(\overline{\theta}_k, k) \tag{5}$$

$\overline{\theta}_k$ is average posterior of the parameter in the kth model. The better model has smaller deviance value.

5. Draw a thematic map that plots the distribution of economic growth of the regency to provinces.

6. Make an interpretation of the results of the modeling; then write conclusions and suggestions.

3. Characteristics of research variable

A hierarchical linear model is a regression modeling that can accommodate a hierarchical data structure. The predictor variables were prepared at all predefined levels, while the response variable was measured at the lowest level [10, 12]. The hierarchical structure model could be established by two levels of models, i.e., the micro models (model at the first level) and macro models (model at the second level). Micro models could be in the form of distribution of data in the first level or the regression model between the observed response and predictor in the first level. Macro models, on the other hand, are usually as the regression model between the parameter of the distribution or the regression coefficients from micro models and the predictor variables measured on the second level [13]. In this case, predictor variables measured in the first level were financial credit distributions, i.e., 17 major economic sectors in the regency, while in the macro modeling, variables related to the provincial level were employed. There are six economic sectors that have the greatest contribution among the 17 major economic sectors at the regency level to economic growth, i.e., trade (x_7), manufacturing industry (x_4), construction (x_6), agriculture (x_1), transportation, warehousing and communication (x_9), and accommodation, food, and beverage services (x_8). At the provincial level, for the variable component macro model, on the other hand, they are inflation (w_1), interest rates on loans (w_2), deposits (w_3), and the ratio of the nonperforming loan (NPL) (w_4).

The distribution of the response variable has to be determined in order to build the likelihood distribution which will be applied in both general one-level Bayesian and hierarchical structure Bayesian approach. To do so, the goodness of fit (GOF) test has to be done to check the suitability of the selected hypothetical distribution pattern with the distribution of the observed data. In this study, the null hypothesis of "the response data follow a particular distribution pattern" would be tested to the alternative hypothesis of "the response data do not follow a particular distribution pattern" by using the Anderson-Darling (AD) test [14]. Eq. (6) represents the AD test statistic:

$$W_n^2 = -n - \frac{1}{n}\sum_{j=1}^{n}(2j-1)\left[\log u_j + \log\left(1 - u_{n-j+1}\right)\right], \tag{6}$$

where n is the number of observed sample units and u_j is the cumulative distribution function at the data observations. The null hypothesis would be rejected when W_n^2 is greater than a critical value, c_α [15], calculated as Eq. (7):

$$c_\alpha = a_\alpha * \left(1 + \frac{b_0}{n} + \frac{b_1}{n^2}\right), \tag{7}$$

where at the significance level $\alpha = 5\%$, the value for $a_\alpha = 0.7514$, $b_0 = -0.795$, and $b_1 = -0.890$ [15]. In this study, the response data was tested whether the pattern was normally distributed or not by using the following hypothesis test.

H_0: The economic growth distribution fits the normal distribution.

H_1: The economic growth distribution is unfit for the normal distribution.

Results of the GOF test by using the AD test show that the economic growth (response variable) of the selected 11 provinces follows the normal distribution. The Bayesian normal-based approach employing the likelihood of normal distribution, therefore, is applicable for this case.

4. Indonesia's economic growth modeling using general one-level Bayesian methods

In the general one-level Bayesian modeling for economic growth, it must begin with the assumption that all regencies in the 11 selected provinces have the same level of economic

Parameter	Mean	MC error	2.50%	Median	97.50%
β_0	5.54400	8.16E−04	5.26400	5.54400	5.825000
β_1	−0.24950	0.001269	−0.66500	−0.24940	0.165100
β_2	0.03708	0.001033	−0.29370	0.03742	0.363600
β_3	0.01071	0.001133	−0.38670	0.01121	0.413800
β_4	−0.02010	0.001788	−0.65760	−0.01853	0.621500
β_5	−0.30450	9.90E−04	−0.61510	−0.30430	0.004251
β_6	0.13530	0.003614	−1.17600	0.13380	1.468000
β_7	0.32110	0.003616	−0.90150	0.32280	1.523000
β_8	0.88960	0.003478	−0.40340	0.88640	2.198000
β_9	−0.36810	0.003282	−1.41100	−0.36580	0.674800
β_{10}	−0.26800	0.002072	−1.00700	−0.26790	0.461900
β_{11}	−0.36240	0.003382	−1.49700	−0.36450	0.802700
β_{12}	0.25370	9.14E−04	−0.06085	0.25490	0.570000
β_{13}	−0.60620	0.002666	−1.58900	−0.60710	0.373700
β_{14}	0.14230	0.001768	−0.48590	0.13990	0.778100
β_{15}	−0.01639	0.004265	−1.53900	−0.01792	1.531000
β_{16}	0.37940	0.002525	−0.51430	0.37870	1.274000
β_{17}	−0.58840	7.76E−04	−0.89230	−0.58780	−0.288300
τ	0.15720	9.57E−05	0.13160	0.15660	0.185300

Table 1. Significance testing parameters of one-level Bayesian model.

growth. All of the 17 variables were employed to model simultaneously and give the general one-level Bayesian model as Eq. (8):

$$\begin{aligned} y = {} & 5.544 - 0.2495x_1 + 0.03708x_2 + 0.01071x_3 - 0.0201x_4 - 0.3045x_5 + 0.1353x_6 \\ & + 0.3211x_7 + 0.8896x_8 - 0.3681x_9 - 0.268x_{10} - 0.3624x_{11} + 0.2537x_{12} - 0.6062x_{13} \\ & + 0.1423x_{14} - 0.01639x_{15} + 0.3794x_{16} - 0.5884x_{17} \end{aligned} \tag{8}$$

The next step is to test the parameter significance of this one-level Bayesian model using credible intervals. If the credible interval does not hold zero, then the estimated parameter is significant. The result shown in **Table 1** says that the intercept and the financial credit distribution of international agencies and other extra-national agencies sector to total loans (x_{17}) have a significant influence to their economic growth, but the other 16 variables are insignificant. The insignificance of the 16 variables means that the contribution of these 16 variables is not statistically influential enough for economic growth in each regency, but those sectors cannot be interpreted that they should not be implemented in every regency to support their economic growth. This insignificance can be caused by the random nature of each sector's activities among regions, where, naturally, it should be varied locally, but in this modeling, it is treated and considered to be all the same and global for all regions, to the response variable.

5. Indonesia's economic growth modeling using hierarchical structure Bayesian methods

Two regression models would be established in this hierarchical structure Bayesian approach, i.e., a regression model for the micro model (first level) and macro model (second level), respectively. The regression model in the first level will use 17 variables, and it has to estimate 198 parameters, while the regression model in the second level will use 4 variables, and therefore, it has to estimate 90 parameters. **Table 2** shows six estimated parameters of 18 regression coefficients in micro models for selected six provinces.

Parameter	Aceh	West Java	Central Java	East Java	South Sulawesi	Southeast Sulawesi
β_0	4.0630	4.3330	4.92000	3.0330	6.3470	4.741
β_1	−0.1328*	−1.0220*	0.28610*	−0.0955*	3.3370*	−2352.000
β_2	−1.9210*	0.0192*	−0.00404*	−0.2382*	0.0821*	−981.200
β_3	−37.4500*	−0.8488*	0.86100*	−1.5820	−14.8400*	−39.940
β_4	−174.6000	−0.3119*	−0.23360*	0.3850*	3.0480*	−1023.000
β_5	8.6950*	−0.3027	0.40700*	0.9209*	−0.5125*	−340.000

*The estimated parameter was not significant at $\alpha = 5\%$.

Table 2. Six estimated parameters of 18 regression coefficients in micro models for selected six provinces.

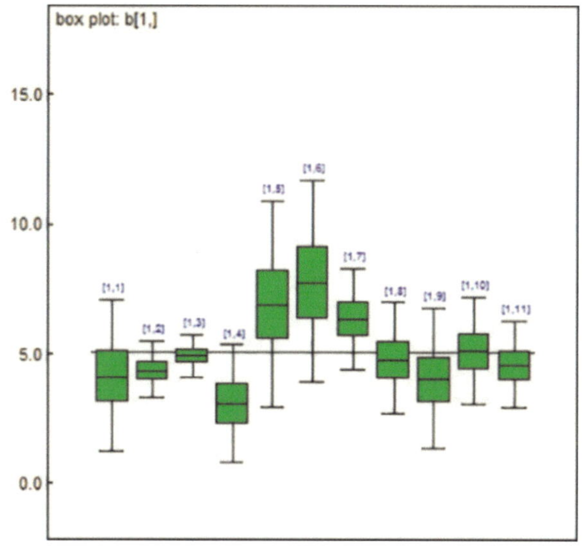

Figure 4. Boxplot of intercept micro models.

Table 2 demonstrates that each estimated β_i, $i = 0, 1, \ldots, 5$ is treated as variables, i.e., the values of β_i among provinces were different. This also applies to β_i, $i = 6, 7, \ldots, 17.\beta_i$, $i = 6, 7, \ldots, 17$. As an example, in the intercept coefficient, the lowest belonged to the East Java Province, while the greatest value belonged to Papua Province. These intercept variations of selected 11 provinces in micro models are presented as boxplot in **Figure 4**. This fluctuation of these parameters would be explained by regressing these parameters to the four covariates in the second level. This has to be done to find out the different effects of their different local policies in implementing their provincial regulations when it is viewed from differences of parameter values [12, 16]. This stage of regression is applied to each random resulted regression parameter of the first level to the covariate at the second level. **Table 3**, as an example, shows only 6 of 18 regressions of macro model. Combining this cross-level interaction hierarchically between micro and macro models, the model of Aceh province, for example, for the randomly intercept only, can be written as Eq. (9):

$$\begin{aligned} y_1 = {} & (5.059 + 1.037w_1 + 0.2406w_2 - 1.605w_3 + 1.153w_4) - 0.1328x_1 \\ & - 1.921x_2 - 37.45x_3 - 174.6x_4 + 8.695x_5 - 11.91x_6 - 0.5904x_7 \\ & - 2.461x_8 - 74.23x_9 + 111.7x_{10} + 238.3x_{11} - 115.1x_{12} - 50.71x_{13} \\ & + 22.8x_{14} + 11.58x_{15} + 0.1152x_{16} - 1.606x_{17} \end{aligned} \tag{9}$$

From the example of a hierarchical model for Aceh, a hierarchical structure model can demonstrate its superiority in presenting a new model as a hierarchical cross-level interaction through the modeling of the slope of micro models. This model can describe the differences in economic growth between different provinces even though they have characteristics of regencies with almost perfect similarities. In this case, the role of provincial characteristics is as an activator variable in relation to the regency's economic growth rate. The interpretation, therefore, could

Parameter in micro model	γ_0	γ_1	γ_2	γ_3	γ_4
β_0	5.0590	1.0370*	0.2406*	−1.6050*	1.1530*
β_1	4.7120E+04	−4.6160E+04	365.3001	9.610E+03	−6.1450E+04
β_2	9.9490E+05	9.6220E+05	−7.2320E+05	−2.9160E+05	−7.3050E+05
β_3	−7.4470E+04	−2.37E+05	1.5040E+05	7.2280E+04	−5.163E+03
β_4	−4.0340E+05	3.4290E+05	4.6780E+04	−3.5150E+04	4.8380E+05
β_5	1.0480E+04	−1,28E+05	3.4670E+04	2.9410E+04	−5.1220E+04

*The estimated parameter was not significant at $\alpha = 5\%$.

Table 3. Summary of parameter estimation macro model regression.

be derived from the micro models adapted to the characteristics of each province. In addition, the creation of predictors by adding depth to the hierarchical level will be more adaptive in capturing real phenomena in the field.

The results in the first line of **Table 2** and **Figure 4** showed that the intercept of micro models varies among the provinces. It is due to the significant effect of the province characteristics as shown by the first line of **Table 3**. All of the estimated parameters of the covariate w in the second level, γ_t, $t = 0, 1, 2, \ldots, q$, are significant, except for γ_0. This means that variables in the second level, inflation (w_1), interest rates on loans (w_2), and NPL ratio (w_4), are affecting the different shifts in economic growth in each regency. Interpretation for parameters other than intercepts can be done in the same way, namely, by substituting the results of parameter estimates at level two in **Table 3** of the second row into the first-level model.

6. The best model selection

Modeling of economic growth in Indonesia in this study is done using two methods, the general one-level Bayesian and the two-level hierarchical structure Bayesian models. These two models would be compared to see which model is a more representative model to economic growth. The main point of view that needs to be highlighted in the modeling differences is that in general one-level Bayesian modeling, all of the characteristics at the provincial level are ignored and only the characteristics in the Regency are considered. In this modeling view, the economic growth in all regencies was, therefore, treated equally. The Bayesian hierarchical structure modeling, on the other hand, was smartly joining the

Model	DIC
General one-level Bayesian	1351.360
Hierarchical structure Bayesian	916.490

Table 4. Goodness-of-fit model.

characteristics at the provincial level and at the regency level. Here, the economic growth could be explained as a cross-level interaction hierarchically through the modeling of parameters of micro models to the province characteristics as its covariates. The criteria used to select the best model are the value of DIC. Based on the smaller DIC in **Table 4**, the hierarchical structure Bayesian model was better than the general one-level Bayesian model.

7. Thematic map of economic growth in Indonesia

The economic growth of each regency based on the one-level Bayesian method can be seen in **Figure 5**. Each color in **Figure 5** represents economic growth in a regency. Color code 64 is a color code for the regency that is not included in the modeling. The higher the economic growth in a region, the greater the color code. Nabire has the highest economic growth in 2015 among other regencies in Indonesia, i.e., 9.51%, so the color code for Nabire is 255.

Furthermore, the thematic map of economic growth of each regency based on the hierarchical Bayesian method shown in **Figure 6**, where the color codes for regencies that are not included modeling is code 83. Like a thematic map of economic growth based on the one-level Bayesian method, these maps also show that the higher the economic growth in a region, the greater the color code.

Based on **Figures 5** and **6**, the difference between economic growth modeling using the global one-level Bayesian method and the hierarchical Bayesian method is easily seen. The difference is the color in **Figure 5** only influenced by covariates of regencies, whereas the difference in

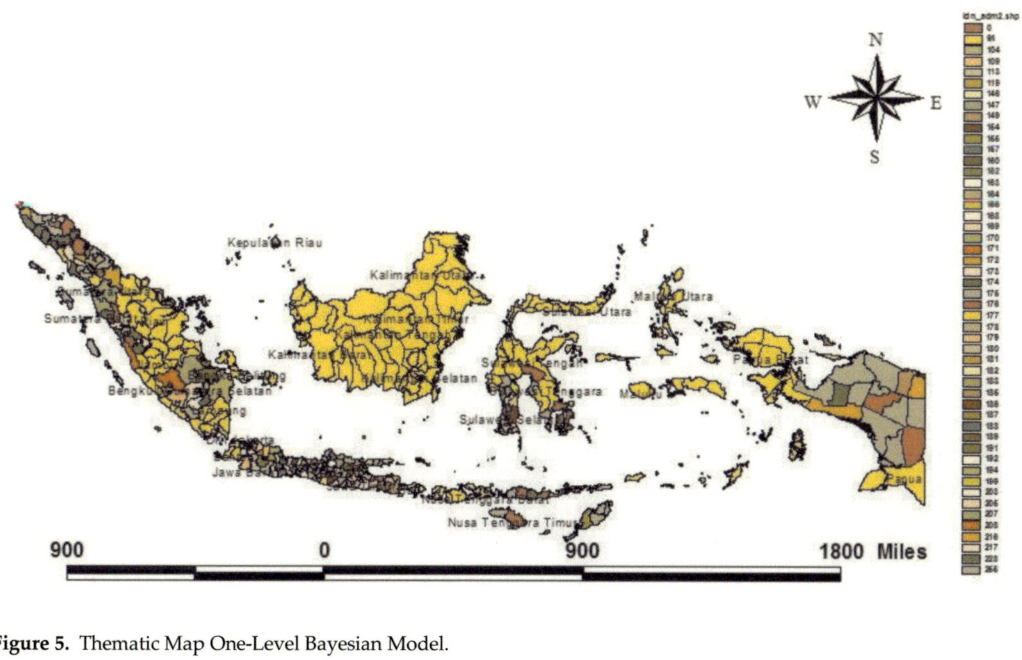

Figure 5. Thematic Map One-Level Bayesian Model.

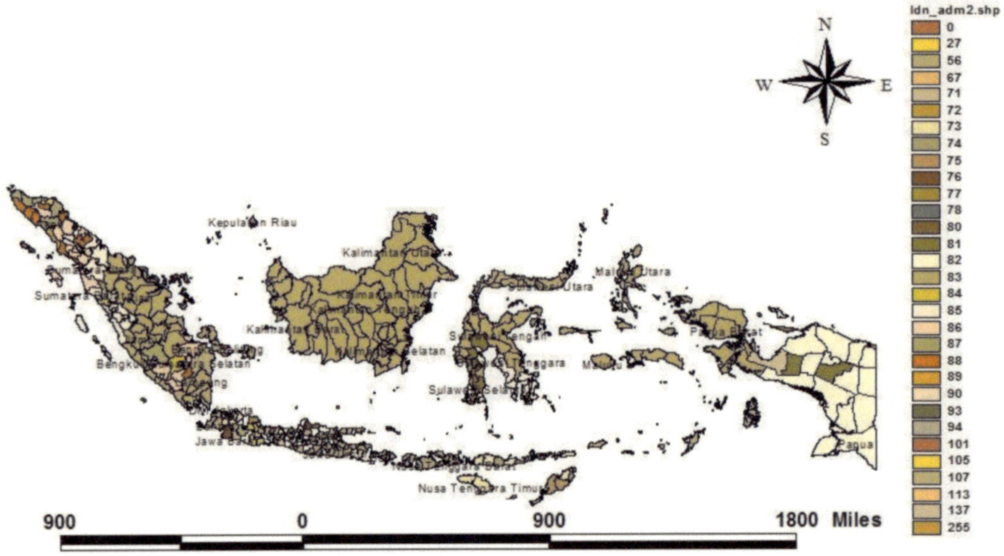

Figure 6. Thematic Map Hierarchical Bayesian Model.

color in **Figure 6** is due to the collaboration and interaction between covariates in each regency and province. In addition to this, collaboration and interaction of the characteristics of the regency and province also affect the color difference in maps. Hierarchical Bayesian model is better in representing the economic growth in each regency in Indonesia, as has been discussed in Section 6. **Figure 6** looks more clear in describing and representing Indonesia's economic growth in 2015.

8. Conclusion

Some conclusions could be gathered, i.e., (i) the economic growth model based on financial credit distribution in Indonesia generally follows the normal distribution pattern, (ii) it would be more appropriate to be modeled using the hierarchical Bayesian than using a global one-level Bayesian method, and (iii) the results of hierarchical Bayesian modeling can also be seen as a significant influence on the regression coefficients that describe a cross-level interaction of the regency and provincial characteristics. The influence of the regency characteristics, therefore, cannot be generalized, so that the regency characteristics should be fitted to the province characteristics.

There were also some recommendations to be given, i.e., (i) the local government and Bank Indonesia should focus on addressing issues of inequality of economic growth in Indonesia, especially in areas with slow rate of economic growth; and (ii) it was necessary to develop a new method that (a) was capable to include the provinces with the regency number of less than 17 variables in the model and (b) was able to model the different distribution patterns of economic growth in the different regions into the generalized hierarchical Bayesian model.

Acknowledgements

We would like to express our gratitude to the anonymous reviewers for the criticism and suggestions for the improvement of this research and to IntechOpen for sponsoring the publication of this article.

Author details

Nur Iriawan* and Septia Devi Prihastuti Yasmirullah

*Address all correspondence to: nur_i@statistika.its.ac.id

Department of Statistics, Faculty of Mathematics, Computing, and Data Science, Institut Teknologi Sepuluh Nopember, Indonesia

References

[1] Todaro MP, Smith SC. Economic Establishment (Originally: Pembangunan Ekonomi). Jakarta: Erlangga; 2006

[2] Statistics Indonesia. Gross Domestic Product. Available from: https://www.bps.go.id/Subjek/view/id/11

[3] United Nations Development Programme (UNDP). Converging Development Agendas: Nawacita, RPJMN, and SDGs. Available from: http://www.id.undp.org/content/dam/indonesia/2015/doc/publication/ConvFinal-En.pdf

[4] Otoritas Jasa Keuangan. Potential Economic Growth. Available from: http://www.ojk.go.id/id/berita-dan-kegiatan/publikasi/Documents/Pages/Potensi-Pertumbuhan-Ekonomi-ditinjau-dari-Penyaluran-Kredit-Perbankan-Kepada-Sektor-Prioritas/Kajian%20Kredit%20-%20Pertumbuhan%20Eko%20(final).pdf

[5] Levine R. Legal environment, bank, and long-run economic growth. Journal of Money, Credit and Banking. 1998;**30**:596-613. DOI: 10.2307/2601259

[6] Rajan R, Zingales L. Financial dependence and growth. The American Economic Review. 1998;**55**:559-586. DOI: 10.3386/w5758

[7] Soedarmono W, Hasan I, Arsyad N. Non-linearity in the finance-growth nexus: Evidence from Indonesia. Working Paper. 2015. pp. 2-17. DOI: 10.1016/j.inteco.2016.11.003

[8] Raudenbush SW, Bryk AS. Hierarchical Linear Models. Thousand Oaks: Sage Publications; 2002

[9] Yasmirullah SDP, Iriawan N, Sipayung FR. An economic growth model based on financial credits distribution to the government economy priority sectors of each regency in Indonesia using hierarchical Bayesian method. AIP Conference Proceedings. 2017;**1905**. DOI: 10.1063/1.5012264. ISBN: 978-0-7354-1595-9

[10] Boldstad WM. Introduction to Bayesian Statistics. New Jersey: Wiley; 2007

[11] Ntzoufras L. Bayesian Modeling Using WinBUGS. New Jersey: John Wiley & Sons Inc; 2009

[12] Ismartini P, Iriawan N, Setiawan S, Ulama BSS. Toward a hierarchical Bayesian framework for modelling the effect of regional diversity on household expenditure. Journal of Mathematics and Statistics. 2012;**8**:283-291

[13] Iriawan N. Modelling and Data-Driven Analysis. Vol. I. Surabaya: ITS Press; 2012

[14] Anderson T, Darling D. Asymtotic theory of certain goodness of fit criteria based on stochastic process. The Annals of Mathematical Statistics. 1952;**23**(2). DOI: 10.1214/aoms/1177729437

[15] Ang A, Tang W. Probability Concepts in Engineering. New York: Wiley; 2007. pp. 289-396

[16] Wirawati I, Iriawan N, Irhamah. Bayesian hierarchical random intercept model based on three parameter gamma distribution. Journal of Physics: Conference Series. 2017;**855**: 012061. DOI: 10.1088/1742-6596/855/1/012061

Bayesian Networks for Decision-Making and Causal Analysis under Uncertainty in Aviation

Rosa Maria Arnaldo Valdés,
V. Fernando Gómez Comendador,
Alvaro Rodriguez Sanz, Eduardo Sanchez Ayra,
Javier Alberto Pérez Castán and Luis Perez Sanz

Additional information is available at the end of the chapter

http://dx.doi.org/10.5772/intechopen.79916

Abstract

Most decisions in aviation regarding systems and operation are currently taken under uncertainty, relaying in limited measurable information, and with little assistance of formal methods and tools to help decision makers to cope with all those uncertainties. This chapter illustrates how Bayesian analysis can constitute a systematic approach for dealing with uncertainties in aviation and air transport. The chapter addresses the three main ways in which Bayesian networks are currently employed for scientific or regulatory decision-making purposes in the aviation industry, depending on the extent to which decision makers rely totally or partially on formal methods. These three alternatives are illustrated with three aviation case studies that reflect research work carried out by the authors.

Keywords: Bayesian networks, prediction, classification, risk, anomaly detection, causal modelling, uncertainty

1. Introduction

Technical and managerial decision-making is a critical process in any industry and any business. Information is a fundamental cornerstone in the decision process, although sometimes its availability and quality are limited or affected by uncertainty.

Uncertainty refers to the stochastic behaviour of a system and to the uncertain values of the parameters that describe it. Most decisions in aviation systems and operation are currently taken under the assumption that the values of the parameters describing the system performance are equal to their estimates. However, this postulation is only valid as long as there are sufficient data or precise expertise for an accurate estimation of the system parameters. This is not the case in many occasions, particularly when the system, product or process is new and limited measurable information about its performance is accessible. Additionally, in many occasions, decision makers in aviation do not count with the assistance of formal methods and tools to help them cope with all those uncertainties in the decision-making process, particularly when it is necessary to evaluate risks or perform causal analysis.

A systematic approach for dealing with uncertainties in aviation and air transport is possible through Bayesian analysis. Bayesian Networks (BNs) have been broadly applied to decision-making problems in a wide variety of fields because they combine the benefits of formal probabilistic methods, understandable easily visual form, and efficient computational tools when exploring consequences and risks.

In this chapter, we revise the advantages of applying BNs to aviation and air transport decision-making problems in environments affected by uncertainty. We characterise typical problems existing in aviation and air transport, which could benefit from this systematisation; and describe recent research work carried out in this field. More particularly, the chapter illustrates works performed by the authors regarding:

i. How Bayesian reasoning can support an integrated methodology to assess and evaluate compliance with system safety goals and requirements when there is uncertainty in the assessment of systems performances.

ii. How Bayesian networks can be used to evaluate the risk of runway excursion at an airport and decide whether an airline will be authorised to operate at that airport vis-a-vis of the operational risk.

iii. How causal analysis through a BN can be used to understand the interdependencies between factors influencing performance and delay (drivers and predictors) at busy airports.

2. Bayesian networks for decision-making in aviation

In general, we may consider three main ways that Bayesian networks are currently employed in causal and risk analysis for scientific or regulatory decision-making purposes in the aviation industry. While in general decision makers prefer to rely on formal infrastructures to back up its decisions, the extent up to what they totally or only partially trust on the formal methods is in the origin of this triple approach.

i. In the first way, the Bayesian reasoning assumes the entire process of evaluation and decision. In this case, the Bayesian approach applies to all the phases and steps in the process and estimations, and decisions respond to an overall Bayesian framework. Typical decision problems normally tackled with this approach addresses questions such as:

- Should a company be allowed to operate at a new airport?
- Does an on-board system satisfy the prescribed safety objectives?
- Should a new aircraft model be certified and allowed to fly?

Those in favour of this approach sustain that Bayesian reasoning is able to provide such an all-inclusive and formal scheme to arrive at decisions, and that applying a scientifically homogenous approach to all the phases of the decision-making process guarantee coherent, objective and solid decisions. Those against this approach claim that with this approach, the Bayesian analyst is put in charge and takes over the entire process and endeavour. Although widely applied in other industries, its use is still rare in aviation.

ii. In the second option, Bayesian methods can be used just to estimate probability distributions. In this case, Bayesian analysis is still a central piece of the decision-making process, although it is not anymore in charge of the whole process. Typical questions addressed by this application of Bayesian methods are:

- What are the odds of an aircraft suffering a runway overshoot?
- What is the probability that a flight will experience a delay?
- What is the probability that passengers will lose their flight?

In this case, Bayesian analyst furnishes the quantities and probability distributions that will help managers to take informed decisions but will not condition their decision, which might be influenced by other factors. Therefore, the decision process is formally isolated from the Bayesian analysis.

iii. At the opposite end, Bayesian methods can be used to select or parameterise input distributions for a probabilistic model. In this case, neither the model nor the decision process relay on the Bayesian methods. Bayesian analysis is reduced at a basic role and is used to estimate the input parameters to many complex models, instead of answering questions directly. This is the simplest application of Bayesian methods in a decision-making process, and it normally constitutes the first application when Bayesian methods are introduced in a new industry.

This application is of particular interest when there are too little data available to sustain statistical analysis, and the only source of available information should be obtained from expert knowledge. Most decisions in aviation are taken under the assumption that the values of the parameters describing the system performance are equal to their estimates, which is only valid as long as there are sufficient data or precise expertise for an accurate estimation of the system parameters. It is not the case in many situations, particularly when the system, product or process is new and tiny measurable information about its performances is accessible. In these cases, BNs represent a framework of causal factors linked by conditional probabilities, which are elicited from aviation experts. Best-expert estimates will use the best available and accessible data.

Typical questions answered by this approach are:

- What is the distribution of partial and total failures of an aircraft component?
- What is the in-service time of an aircraft component?
- What is the uncertainty about the probability of a critical event? and
- How can we characterise uncertainty about the aircraft trajectories or delays?

When talking about the different areas of aviation, the application of Bayesian networks is not homogeneous. Several respected research groups and authors have initiated the application of BNs in aviation. In fact, literature nowadays is wide enough to support reviews as the ones recently performed by Broker in [1] or Roelen in [2], about BN applications for aviation risk estimation.

Aviation safety and risk analysis are by far the domain where more BN applications can be found. A thoughtful revision shows that this technique is particularly useful to provide additional insights into problems of "low probability-high consequence," such as the aviation safety domain where events occur very infrequently.

- In [3], Bayesian Belief Networks are applied to model a number of safety defensive barriers in Air Traffic Control environment from airspace design, through tactical control, and from the operation of aircraft safety net features to a potential accident.
- In [4], Luxhoj and Coit used Bayesian networks to model a certain aircraft accident type known as Controlled Flight Into Terrain (CFIT).
- In [5], the authors develop causal models for air traffic using "event sequence diagrams, fault-trees and Bayesian belief nets linked to form a homogeneous mathematical model suitable as a tool to analyse causal chains and quantify risks…".
- Some authors [6] have developed an inclusive aviation safety model to evaluate management decisions potential impact.
- Ref. [7] introduces a BN for the evaluation of flight crew performance, and Delphi technique to complement data from accident reports
- Problems at very low level of detail regarding safety in operational issues have also benefited from the application of Bayesian methods [8].
- Reducing aviation safety risk is a matter of concern also for NASA, who focuses on the reasoning of selecting Object-Oriented Bayesian Networks (OOBN) as the technique and commercial software for the accident modelling [9].
- In [10], a BN analysis model is established by using 10 years of flight crew members' error data in China civil aviation incidents to analyse the probability distribution of flight crew members' errors in civil aviation incidents analysis.
- Several models have attempted to explain various factors influencing aeronautical accidents: human, organisational, environmental and airport infrastructure factors. The model by [11] permits to evaluate the influence of these factors and identify the dependence and relationship among them.

- A very initial attempt to assess aviation security can be found at [12], which addresses the evaluation and mitigation of security risks in the aviation domain and realises a multi-dimensional approach of complex systems.
- Bayesian networks are capable of providing real-time safety monitoring functionalities, like those in [13] that integrates automatic video analysis algorithms and Bayesian models to detect anomalous behaviours of ATCs and spatiotemporal details about how errors due to fatigue and distractions eventually lead to near-ground incidents/accidents.
- In [14], Arnaldo et al. used Bayesian inference and hierarchical structures to predict aircraft safety incidents.

The second domain where more BNs can be found is operational analysis, particularly delays optimisation. BNs represent a paradigm shift in the study of aviation delays because they have a structure that is machine-learned from data and do not require assumptions about "causal" patterns; they can produce estimates even in situations with sparse or limited data, and they can be used well in advance of the actual flight, as they can predict based on only partial evidence.

- In [15], the random characteristics of civil aviation safety risk are analysed based on flight delays, using a BN to build an aviation operation safety-assessment model based on flight delay.
- The propagation of micro-level causes to create system-level patterns of delay, a problem difficult to assess by traditional methods, has been assessed with BNs to investigate and visualise propagation of delays among airports, demonstrating greater predictive accuracy than using linear regression [16].
- In [17], a new Bayesian Network algorithm, Negotiating Method with Competition and Redundancy (NMCR), demonstrate excellent performances in estimating of arrival flight delay, especially in flight chains mainly operated in China.
- The NextGen Advanced Concepts and Technology Development Group of the FAA (Federal Aviation Administration) have tackled this problem by developing Bayesian Networks for Departure Delay Prediction [18].
- The aviation supply chain has also been modelled through Bayesian networks to minimise delays causing factors [19].
- Another relevant case on airport delay analysis can be found in [20]. This chapter develops a functional analysis of the operations that represent the aircraft flow through the airport airspace system. By considering the accumulated delay across the different processes and its evolution, different metrics are proposed to evaluate the system's state and its ability to ensure an appropriate aircraft flow in terms of time saturation.

Another area that has received attention from Bayesian experts is the modelling of airline risk considering reliability data, maintainability data and management data.

- Some attempts have been made to approach software health management based on a rigorous Bayesian formulation to monitor the behaviour of software and operating system, to perform probabilistic diagnosis, and to provide information about the most likely root causes of a failure or software problem. Three realistic scenarios from an aircraft control system were considered: (1) aircraft system-based faults, (2) signal handling faults, and (3) navigation faults due to inertial measurement unit (IMU) failure or compromised Global Positioning System (GPS) integrity [21].

- Ref. [22] covers the construction of a probabilistic risk analysis model for the jet engines manufacturing process, based on BN coupled to a bow-tie diagram. It considers the effects of human, software and calibration reliability to identify critical risk factors in this process. The application of this methodology to a particular jet engine manufacturing process is presented to demonstrate the viability of the proposed approach

- BN has also been designed for fault detection and isolation schemes to detect the onset of adverse events during operations of complex systems, such as aircraft and industrial processes [23].

- Another relevant work on fault diagnosis is the one by [24] to study automatic fault diagnosis of IFSD (in-flight shutdown).

- In the area of maintenance, BNs are also applied for improving Human reliability analysis (HRA) in visual inspection [25].

Finally, one of the most attractive probabilistic modelling framework extensions of Bayesian Networks for working under uncertainties from a temporal perspective, Dynamic Bayesian Networks (DBNs), has also had some applications in aviation.

- DBNs have been used to model abnormal changes in environment's data at a given time, which may cause a trailing chain effect on data of all related environment variables in current and consecutive time slices.

- In [26], an algorithm is proposed for pilot error detection, using DBNs as the modelling framework for learning and detecting anomalous data, based on the actions of an aircraft pilot, and a flight simulator is created for running the experiments. The proposed anomaly detection algorithm has achieved good results in detecting pilot errors and effects on the whole system.

- Another application to dynamic operational problems can be found in [27], where the variables which affect the Helicopter's real-time aviation decision process are represented on Structure Variable Discrete Dynamic Bayesian Network, building up a model that could be used in real-time aviation decision process in perpetual variational air combat.

- From a point of view, less operational and more economical, BNs also help the aviation industry and dynamically recommend airline managers relevant contents based on predicting passengers' choice to optimise the loyalty.

The remaining sections of the document illustrate the application of each one of the three options, enumerated at the beginning of this section, through three aviation case studies that reflect research works carried out by the authors.

http://dx.doi.org/10.5772/intechopen.79916

3. Case study 1: Bayesian framework for safety compliance assessment and acceptance under uncertainty

In [28], we present a good example where Bayesian reasoning assumes the entire process of evaluation and decision. This work presents an integrated methodology, based on Bayesian inference, to assess and evaluate compliance with system safety goals and requirements when there is uncertainty in the assessment of systems performances.

Compliance assessment process is addressed in this work as a Bayesian decision problem:

$$B = \langle A, N, P, W, U \rangle, \tag{1}$$

where

- A states for the decision maker actions space, a_i, $A = \{a_1, a_2, \ldots a_n\}$
- N represents the space of possible "states of nature", i.e. magnitudes about which there is uncertainty, $N = \{Ns_1, Ns_2\} = \{Cs, \overline{Cs}\}$
- P represents the space of uncertainties about the state of nature of the system, $P = \{P(Ns_1), P(Ns_2),\} = \{P(Cs\,|D, I), P(\overline{Cs}\,|D, I)\}$
- W represents the set of decision outcomes, $W = \{W_{11}, W_{12}, \ldots, W_{ij}, \ldots, W_{nm}\}$
- U represents the set of utility functions, $U = \{u_{11}, u_{12}, \ldots, u_{ij}, \ldots, W_{nm}\}$

Each combination $(a_i, Ns_i) \in C = A\ x\ N$ determines a consequence of a course of action for the decision maker. The utility function $u_{ij}(c)$ defines the predilections of the decision maker on a course of action a_i for a system with a state of safety complianceNs_j.

The overall process of safety compliance assessment is addressed through a Bayesian approach as illustrated in **Figure 1**. The rectangle at the left-hand part of the figure represents a decision node, which displays the three potential actions, a_i, which the decision maker can take as a result of the safety compliance process:

- a_1 - Judge the system compliant;
- a_2 - Judge the system as non-compliant; or
- a_3 - Judge the information insufficient.

The circles denote random nodes, which represent the "states of nature", that is, the actual state of system compliance, Ns_j, where

- $Ns_1 = C_s$; $Ns_2 = \overline{C_s}$

Being the notation of C_s the event that the system is actually compliant, whereas $\overline{C_s}$ denotes the event that the system is not actually compliant. The uncertainties in the states of nature P_j

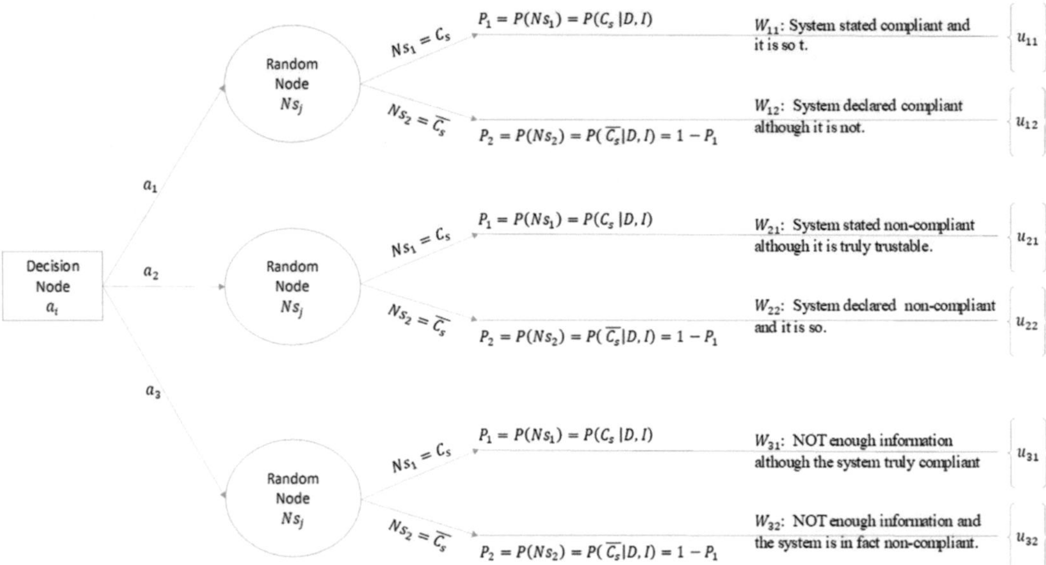

Figure 1. Bayesian decision tree for safety acceptance of a system.

are provided by the Bayesian estimation process. The belief or uncertainty about the compliance state of the system C_s is dependent on the data D and information I available.

- $P_1 = P(Ns_1) = P(C_s \mid D, I)$;
- $P_2 = P(Ns_2) = P(\overline{C_s} \mid D, I) = 1 - P_1$

Each of the branches of the tree represents the set of possible (unpredictable) outcomes W_{ij} that can occur under each action taken by the decision maker. The six possible outcomes, in this case, correspond to:

- W_{11}: The system is stated compliant and it is so;
- W_{12}: The system is declared compliant although it is not;
- W_{21}: The system is stated non-compliant although it is truly trustable;
- W_{22}: The system is declared non-compliant and it is so;
- W_{31}: The decision maker has no enough information although the system truly compliant;
- W_{32}: The decision maker has no enough information and the system is in fact non-compliant.

Safety compliance is assigned a probability of being true, which represents the decision maker uncertainty (or state of knowledge), about its truth or falsity. Namely, the uncertainty on the state of nature of the system compliance considering previous knowledge and information is expressed as: $P(Ns_n) = P(Cs \mid D, I)$, where a proposition D stands for data and I stands for background information. This framework subscribes to the concept that probability is not a frequency, rather a measure of uncertainty, belief or a state of knowledge. That is, probability allows doing plausible reasoning in cases where we cannot reason with certainty.

The result is the predictive probability that the system meets the safety objectives for what it has been designed, considering the envelope of data, knowledge and information gathered from the system during its design, production and operation.

To that aim, compliance assessment is redefined as the determination of the degree of belief in the fulfilment of the applicable failure probability objectives by the candidate system, for all failure conditions N. The whole system is considered compliance if all the λ_n satisfy their pertinent failure safety objective O_n. In this step, the principles of Bayesian inference are applied to improve the estimation of the system/ component rate of failureλ_n.

The conditional probability distribution P$(\lambda_n \,|D, I)$ describes then the uncertainty in the parameter under study (λ_n) considering new events D and the prior understanding of the system I. It represents the sampling distribution of the rate of failure conditional upon the observed data and information and is precisely the form required for decision-making without the need for approximation. It is determined using the Bayes' theorem:

$$P(\lambda_n \,|D, I) = \frac{P(D|\lambda_n, I) \times P(\lambda_n \,|I)}{P(D|I)} \tag{2}$$

where

- $P(\lambda_n \,|D, I)$ corresponds to the posterior distribution. The posterior distribution will the foundation for all inference about the parameter λ_n;
- $P(D|\lambda_n, I)$ corresponds to the likelihood distribution, sometimes referred as sampling;
- $P(\lambda_n \,|I)$ is the prior distribution; and
- $P(D|I)$ is the failure of unconditional or marginal probability D.

Epistemic uncertainty is incorporated through the Prior distribution $P(\lambda_n \,|I)$. It epitomises the degree of belief in model parameters λ_n and defines an initial state of knowledge. Prior distribution can be non-informative or informative. Non-informative priors include very little fundamental info regarding the unknown and facilitates data dominate the posterior distribution. Other terms for non-informative priors are diffuse priors, vague priors, flat priors, formal priors, and reference priors. Informative priors provide essential information about the unknown parameter. Historical data and expert judgement can be incorporated into the prior probability distribution. Although the prior can take the form of any distribution, conjugate priors simplify the evaluation of the previous equation and allow analytical solutions avoiding the use of numerical integration. In practice, the Bayesian approach often leads to intractable integrals and numerical simulation procedures need to be adopted. Normally, due to the complexity of the distributions, the solution of Equation has to be accomplished by numerically Markov Chain Monte-Carlo (MCMC) simulation.

The resulting posterior distribution, $P(\lambda_n \,|D, I)$, stands for updated knowledge about λ_n and is the basis for all inferential statements about λ_n.

The distribution $P(D|\lambda_n, I)$ represents the chance of the data D and model aleatory uncertainties. It represents inefficiencies in the data collection as well as the failure mechanism or the failure model. Likelihood functions commonly used in safety assessment are binomial, Poisson, or exponential ones.

Finally, $P(D|I)$ is just a normalisation constant.

$P(C_{sn}\ |D, I)$ can be inferred from the posterior distributions $P(\lambda_n|D, I)$ through marginalisation of the parameter λ_n, as indicated in the following equation.

$$P(C_{sn}\ |D, I) = \int_{\wedge} P(O_n, \lambda_n\ |D, I).d\lambda =$$

$$\int_O^{O_n} P\ (O_n|\lambda_n)P\ (\lambda_n|D, I).d\lambda_= \int_O^{O_n} P\ (O_n|\lambda_n)\frac{P(D|\lambda_n, I) \times P(\lambda_n\ |I)}{P(D|I)}.d\lambda \quad (3)$$

Eq. (3) computes an average of the model uncertainty integrating the sampling distribution $P\ (O_n|\lambda_n)$ over the posterior distribution $P(\lambda_n\ |I)$. The output is a predictive probability of a failure condition meeting its safety objective.

This Bayesian framework espoused is exemplified over a practical case. This practical case corresponds to a real situation with current hypothesis, requirements and data: a new ANSP initiates the provision of Tower Control and CNS (Communications, Navigation and Surveillance) services at the new international airport of Castellón (Spain).

The service provider is subject to supervision by the National Aeronautical Authority and must demonstrate compliance with applicable safety requirements. At Castellón airport, air navigation service comprises ground-based radio navigation aids, very high-frequency omnidirectional range (VOR), distance measuring equipment (DME), and precision approach and landing aids, instrument landing system (ILS). The functionalities of each of these systems and the applicable requirements are regulated at international level. Providers of air navigation services must prove that their operating procedures and working methods are compliant with the prescriptions and standards of ICAO Annex 10. They must guarantee the accuracy, continuity, availability and integrity, as well as the quality level, of their services.

4. Case study 2: runway excursion

In [29], the authors work on a representative example of the option where Bayesian methods are used to estimate probability distributions. Statistics about commercial aircraft fleet accident produced by Boeing (2012) states that around 37% of the accidents took place during landing and final approach flight phases, and among them, runway excursions accounted for 25% of all accidents. In particular, within the runway excursions, those that are produced by a too long landing (overrun excursion) represent 96%, and the 10-year moving average during 1992–2011 indicates a deteriorating tendency.

This section summarises the work done by the authors to develop a Bayesian model to evaluate the runway overrun risk at a given airport and operational conditions. The model allows comparing the probability of excursion at landing at several runways or airports. The model relates overrun probabilities with possible generating factors, then suggesting the outline of mitigation actions.

The probabilistic influence diagram for runway overrun Bayesian network (see **Figure 2**) is based on the information from safety authorities, operators and manufacturers [30–32]. The network combines expert judgement and data analysed with the aid of the GeNIe SW.

The critical variable chosen as network outcome is "the remaining runway at 80 kt (I), measured in ft", since, as indicated by the FSF SLGs [33], the risk of a runway overrun increases significantly if when there are just 2000 ft. (610 m) of landing distance available (LDA) the aircraft is not decelerated below 80 kt. The nodes in the network account for:

- Relevant Runway. It is a categorical variable: (A).
- Crosswind component at threshold. Unit of measurements is knots: (B).
- Speed of the aircraft which it is discretised to the nearest integer in the avionic: (kt).
- Tailwind component at threshold. Unit of measurements is knots: (C).
- Stabilised/unstabilised state at the approach: (D).
- Maximum reverse thrust, which describes the maximum reverse thrust is applied during ground roll. It is measured in seconds: (E).
- Autobrake state at landing, which has three values: low, medium, and no autobrake: (F).
- Difference between the Indicated AirSpeed (IAS) and the Final Approach Speed (Vapp): (G)

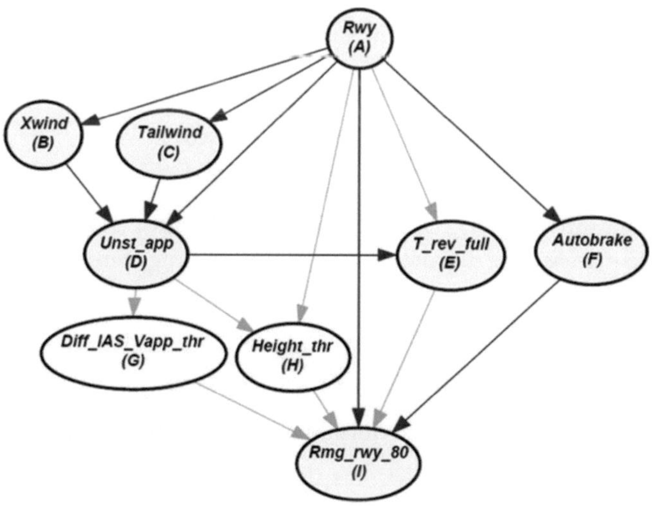

Figure 2. BN for overrun events.

- Aircraft height at threshold, measured in feet (ft): (H).

The safety issue analysed in this work is among the group of most frequently reported accident/incident types all over the world, and it is considered as a big threat to aviation safety. Runway excursions take place with very low frequency, but their consequences may be quite severe. Very low probabilities of occurrence are an added challenge for a risk analyst. Reducing landing overruns is a priority for international aviation organisations that are actively investigating and proposing safety strategies to contain this risk.

The work carried out by the authors in this study uses public information provided by safety agencies, operators and manufacturers; as well as expert judgement and data to create an influence diagram and a probabilistic model.

The model is illustrated with a case study in which three runways are benchmarked in terms of runway excursion risk. The critical event considered to evaluate the risk of runway excursion was the probability that the aircraft not being below 80 kts when just 2000 ft. (610 m) of LDA remains Pr (I < 2000). The case study is a representative of the decision problems, and airline has to cope with when opening new routes and evaluating operation at new airports or with new fleet. To illustrate the usability of the model and its benefits, the case study uncovered the following issues:

- For this specific case study, the Bayesian network and the supporting data allow discarding correlation between cross and tailwind components.
- Although in general, landing with windy, both crosswind and tailwind components, increases the probability of unstabilised approach, however, tailwind influence is not so determinant at runways 2.
- The variables with the toughest effect on the lasting runway at 80 kt were:
    - **i.** the LDA, available landing distance,
    - **ii.** the used of the autobrake system, and
    - **iii.** the difference between the Vapp and the IAS at the threshold.
- Height at the threshold and maximum reverse thrust variables does have a minor effect on the risk of excursions at the three compared runways.
- The network faithfully reflects operational aspects the propensity to pitch down prior to the threshold to increase the distance available for landing, commonly known as "ducking under" effect.
- The probability of slowing the aircraft at 80 kt in the last 2000 ft. of the runway rises as wind, both components crosswind and tail, increase, except for runway 2.
- Crosswind results are coherent with normal operations. With a severe crosswind, the use of the autobrake system is recommended, since it is more difficult to control and decelerate the aircraft.
- Unstabilised approaches are prone to the most hazardous conditions.

- Longer periods of maximum reverse thrust operation, favour reduction of remaining runway at 80 kt, and consequently have a negative effect on risk of runway excursion. Prolonged operation of the maximum reverse thrust may indicate difficulties to decelerate the aircraft during the ground roll. This variable could then be used as a proxy for runway excursion risk by the airlines Flight Data Monitoring (FDM) teams.
- Runway excursion risk increases with longer operation of reverse thrust, which might be an indicator of difficulties to slow down during the ground run. Accordingly, it is recommended to consider this variable as a precursor of runway excursion risk, and closely monitored it in the Airline's Flight Data Monitoring (FDM) programs.

5. Case study 3: airport operation uncertainty characterisation

In [34], the authors analysed the aircraft flow through the Airport focusing on the airspace/airside integrated operations and characterising the different temporal aircraft operation milestones through the airport based on an aircraft flow's Business Process Model and Airport Collaborative Decision-Making methodology. Probability distributions of the factors influencing aircraft processes are estimated, as well as conditional probability relationship among them. The work turned up in a Bayesian network, which manages uncertainties in the aircraft operating times at the airport. This case study constitutes a representative example of the third manner Bayesian networks are currently employed decision-making purposes in the aviation industry.

The work is based on the collection and analysis of nearly 34,000 turnaround operations at the Adolfo Suárez Madrid-Barajas Airport and concluded with several lessons learned regarding the characterisation of delay propagation, time saturation, uncertainty precursors and system recovery.

The BN structure is represented in **Figure 3** and the network variables. It was organised in different layers attending to the nature of the data to facilitate the understanding of the causal relationships among influence parameters. Colours in **Figure 3** represent the different BN layers.

- Nodes 1–5 refer to meteorological conditions.
- Nodes 6–13 count for variables regarding the arrival airspace: timestamps and congestion metrics (throughput, queues and holdings).
- Nodes 14–15, 26 and 38–39 refers to the airport infrastructure.
- Nodes 16, 22–25 and 40 account for the operator, aircraft, route and flight data.
- Nodes 17–21, 27–37 and 41–42 include data about airside operational times and flight regulations
- Nodes 43–49 stand for delay causes.

The probabilistic Bayesian Network is able to predict outbound delays probability distribution given the probability of having different values of the causal control variables, and by setting a

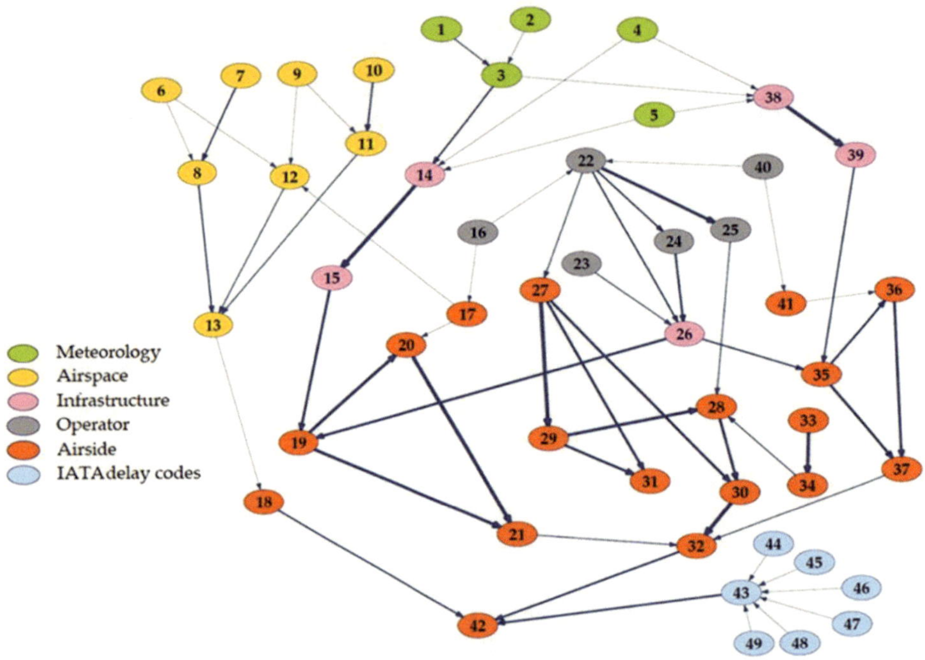

Figure 3. BN model to explain the interdependencies between factors that influence delay performance and system saturation.

target to the output delay, the model provided the optimal configuration for the input nodes. The main outcomes of this work were:

- the statistical characterisation of processes and uncertainty drivers and
- the causal model for uncertainty management (BN).

The case study showed that considering the 34,000 aircraft operations analysed Madrid Airport:

- Arrival delay increases and accumulates its impact over the day, due to network effects.
- However, departure delay does not follow arrival delay's pattern.
- The airport is capable of absorbing a fraction of the arrival delay.
- The main potential drivers for delay include:
  - **i.** time of the day,
  - **ii.** congestion at ASMA,
  - **iii.** weather conditions,
  - **iv.** amount of arrival delay,
  - **v.** scheduled duration of processes,
  - **vi.** runway configuration,

vii. airline business model,

viii. handling agent,

ix. aircraft type,

x. route origin/destination, and

xi. existence of ATFCM regulations.

- Departure delay is highly influenced by the event of longer duration, which at the same time, are the event offering greater possibilities for recovery delays.

6. Conclusions

As stated at the introduction of this chapter, important decisions in aviation systems and operation are currently taken in less than optimal circumstances, under high levels of uncertainty, with only limited amount of data and reliable information, and without the assistance of formal methods and tools.

Based on a thoughtful revision of the available the literature, to determine what domains in aviation and air transport Bayesian Networks applications, the chapter characterises the three main ways that Bayesian networks are currently employed for scientific or regulatory decision-making purposes in the aviation industry, depending on the extent to which decision makers rely totally or partially on formal methods:

i. Bayesian reasoning assumes the entire process of evaluation and decision.

ii. Bayesian methods are used just to estimate probability distributions.

iii. Bayesian methods are used to select or parameterise input distributions for a probabilistic model.

These three alternatives have been illustrated with three case studies that reflect research work carried out by the authors and accounts for the following research questions:

iv. Use of Bayesian decision theory under uncertainty to evaluate compliance with system safety goals and requirements.

v. Runway excursion risks evaluation at an airport, using Bayesian networks to decide about airline initial operation considering the operational risk.

vi. Understand the interdependencies between factors influencing performance and delay (drivers and predictors) at busy airports with using Bayesian networks.

In this work, the authors pretend to highlight the advantages of Bayesian networks as a useful systematic approach to help decision makers to cope with all those uncertainties and difficulties in the decision-making process, particularly when it is necessary to evaluate risks or perform causal analysis.

Author details

Rosa Maria Arnaldo Valdés*, V. Fernando Gómez Comendador, Alvaro Rodriguez Sanz, Eduardo Sanchez Ayra, Javier Alberto Pérez Castán and Luis Perez Sanz

*Address all correspondence to: rosamaria.arnaldo@upm.es

Air Space Systems, Air Transport and Airports Department, School of Aerospace Engineering, Universidad Politecnica de Madrid Plz. Cardenal Cisneros, Madrid, Spain

References

[1] Brooker P. Experts, Bayesian belief networks, rare events and aviation risk estimates. Safety Science. 2011;**49**(8-9):1142-1155

[2] Roelen A. Risk Models of Air Transport. Netherlands: Technische Universiteit Delft; 2008

[3] Neil M, Malcom B, Shaw R. Modelling an air traffic control environment using Bayesian belief netwworks. In: 21st International System Safety Conference; Ottawa. 2003

[4] Luxhoj J, Coit D. Modelling low probability/high consequence events: An aviation safety risk model. In: Annual Reliability and Maintainability Symposium 2006, RAMS '06; 2006

[5] Ale BB, Bellamy L, Cooke R, Goossense L, Hale A. Towards a causal model for air transport safety an ongoing research project. Safety Science. 2006;**44**(8):657-673

[6] Swets L, Zeitlinger. Aviation causal model using Bayesian Belief Nets to quantify management influence. In: Bedford, van Gelder, editors. Safety and Reliability. Tokyo: BALKEMA Publishers; 2003

[7] Wei C, Shuping H. Evaluating Flight Crew Performance by a Bayesian Network Model. Entropy; March 2018

[8] Castilho IX. Fault prediction in aircraft tires using Bayesian networks [MSc in Aerospace Engineering thesis]

[9] Shih AT, Ancel E, Jones SM. Object-oriented Bayesian Networks (OOBN) for Aviation Accident Modeling and Technology Portfolio Impact Assessment, American Society for Engineering Management (ASEM) 33rd International Annual Conference. NASA; Oct 17-20, 2012

[10] Wang Y, Liya J, Mei H. Human factors analysis model of flight crew members based on Bayesian network in China civil aviation incidents. In: The Twelfth COTA International Conference of Transportation Professionals; 2012

[11] Bandeira MCGSP, Correia AR, Martins MR. Method for measuring factors that affect the performance of pilots. Transporte. 2017;**25**(2):156-169

[12] Cioaca C, Boscoianu M. An introduction in the risk modeling of aviation security systems. Mathematics and Computers in Biology, Business and Acoustics. Lupulescu NB, Yordanova S, Mladenov V editors. Barsov, Romania; 2011

[13] Kamala P et al. Automates human performacne monitoring for air traffic control safety through bayesian network modeling and video surveillance. International Journal of Computer Science and Information Technologies. 2015;**6**(5):4392-4396

[14] RMA V, Fernando Gómez Comendador V, Sanz LP, Sanz AR. Prediction of aircraft safety incidents using Bayesian inference and hierarchical structures. Safety Science. 2018;**104**: 216-203

[15] Wang H, Jun G. Bayesian network assessment method for civil aviation safety based on flight delays. Mathematical Problems in Engineering. 2013;**2013**(Article ID 594187):12

[16] Xu N, Donohue G, Laskey KB, Chen CH. Estimation of delay propagation in the national aviation system using Bayesian networks. In: Proceedings of the Fifth USA/Europe Air Traffic management (ATM) R&D Seminar. Baltimore: 2005

[17] Liu Y, Wu H. A remixed Bayesian Netwrok based algorithm for flight delay estimating. International Journal of Pure and Applied Mathematics. 2013;**3**:465-475

[18] Cosmas A. Bayesian networks for departure delay prediction. In: NASA Ames Research Center Airline Operations Workshop; 2015

[19] Yorukoglu M, Kayakutlu G. Bayesian network scenarios to improve the aviation supply chain. In: Proceedings of the World Congress on Engineering, WCE. London; 2011;Vol II

[20] Rodríguez-Sanz Á et al. Analysis of saturarion at the airport and airspace integrated operations. A case study regarding delay indicators and their predictability. In: Twelfth USA/Europe Air Traffic Management Research and Development Seminar (ATM2017); 2017

[21] Schumann J, Mbaya T, Mengshoel O. Software health management with Bayesian networks. Innovations in Systems and Software Engineering. December 2013;**9**(4):271-292

[22] Pereiraa JC, Limaa GBA, Annibal. A bow-tie based risk framework integrated with a Bayesian belief network applied to the probabilistic risk analysis. Brazilian Journal of Operations & Production Management. 2015;**12**:350-359

[23] Gautam B, Mack DL, Kouts XD. Learning Bayesian network structures to augment aircraft diagnostic reference models. IEEE Transactions on Automation Science and Engineering. Jan. 2017;**14**(1)

[24] Huiling C et al. Research on in-flight shutdown fault diagnosis of civil aviation engine based on Bayesian networks. Advanced Materials Research. 2012;**403-408**:1416-1419

[25] Wei Chen SH. Human reliability analysis for visual inspection in aviation maintenance by a Bayesian network approach. In: Transportation Research Record Journal of the Transportation Research Board. December 2014;**2449**(1):105-113

[26] Saada M, Meng Q, Huang T. A novel approach for pilot error detection using dynamic Bayesian networks. Cognitive Neurodynamics. 2014;**8**(3):227-238

[27] Gao PZX. Real-time aviation decision based on structure variable Bayesian network. In: 3rd International Conference on Advanced Computer Theory and Engineering (ICACTE); 2010

[28] Valdés RA, Fernando Gomez Comendador V, Sanz AR, Castán AP, Sanz LP, Ayra ES. Bayesian approach to safety compliance assessment and acceptance under uncertainty for air navigation service providers. Safety Science. Under revision

[29] Sanchez Ayra E. Risk analysis and safety decision-making in commercial air transport operations [PhD thesis]; 2013

[30] Nielsen T, Jensen F. Bayesian Networks and Decision Graphs. Berlin: Springer Publishing Company Inc; 2007

[31] Cowell R, Dawid P, Lauritzen S, Spiegelhalter D. Probabilistic Networks and Expert Systems: Exact Computational Methods for Bayesian Networks. Springer; 2007

[32] Cheng A, Dillard A, Hackler L, Van Der Geest P, Van Es G. Study of Normal Operational Landing Performance on Subsonic, Civil, Narrow-Body Jet Aircraft during Instrument Landing System Approaches. Federal Aviation Administration (FAA); 2007

[33] FSF. Reducing the Risk of Runway Excursions. Flight Safety Foundation (FSF), Virginia U.S.; 2009

[34] Rodríguez-Sanz Á, Comendador FG, Valdés RA, Pérez-Castán JA. Characterization and prediction of the airport operational saturation. Journal of Air Transport Management. 2018;**69**:147-172

Chapter 4

Using Bayesian Networks for Risk Assessment in Healthcare System

Bouchra Zoullouti, Mustapha Amghar and
Sbiti Nawal

Additional information is available at the end of the chapter

http://dx.doi.org/10.5772/intechopen.80464

Abstract

To ensure patient safety, the healthcare service must be of a high quality, safe and effective. This work aims to propose integrated approaches to risk management for a hospital system. To improve patient's safety, we should develop methods where different aspects of risk and type of information are taken into consideration. The first approach is designed for a context where data about risk events are available. It uses Bayesian networks for quantitative risk analysis in the hospital. Bayesian networks provide a framework for presenting causal relationships and enable probabilistic inference among a set of variables. The methodology is used to analyze the patient's safety risk in the operating room, which is a high risk area for adverse event. The second approach uses the fuzzy Bayesian network to model and analyze risk. Fuzzy logic allows using the expert's opinions when quantitative data are lacking and only qualitative or vague statements can be made. This approach provides an actionable model that accurately supports human cognition using linguistic variables. A case study of the patient's safety risk in the operating room is used to illustrate the application of the proposed method.

Keywords: risk assessment, patient's safety, fuzzy Bayesian network, fuzzy logic, Bayesian network

1. Introduction

Medical error is a leading cause of death and injury. Each year, between 210,000 and 440,000 patients who go to the hospital for care suffer from some types of preventable harm that contribute to their death [1]. High error rates with serious consequences are most likely to occur in the operating room [2]. A strong patient's safety culture in the operating room is very important to improve quality and reduce risks of adverse event and medical errors. Thus, a flexible risk analysis technique becomes crucial.

A lot of methods and techniques, such as fault tree analysis (FTA) and failure mode and effect criticality analysis (FMECA), have been used for safety risk analysis in the healthcare system. However, these methods have a limitation when dealing with rare event and complex systems. Khakzad indicated FTA unsuitable for complex problems with its limitation in explicitly representing dependencies of events, updating probabilities, and coping with uncertainties [3], while FMECA does not take into account multiple failure scenarios and causes. Bayesian Network (BN) is a powerful method for risk analysis. In contrast with other classical methods of dependability analysis, Bayesian networks provide a lot of benefits. Some of these benefits are the ability to model complex systems, to make predictions as well as diagnostics, to compute exactly the occurrence probability of an event, to update the calculations according to evidences, to represent multimodal variables, and to help modeling user-friendly by a graphical and compact approach [4].

In this chapter, we propose two methods which can help to assess patient safety in different contexts using Bayesian network.

2. Case of the data availability about risks

In this part, we propose a method for the context of data availability. We will explain how we can use the classical Bayesian network for safety assessment in healthcare system.

2.1. Methodology of risk analysis of the operating room

In the following, a methodology of risk analysis of the operating room using Bayesian networks is proposed. The methodology follows four steps (**Figure 1**) and it is part of continuous improvement process (CIP) [5].

The first step involves determining the aim of the risk assessment process, the description of the problem, and the definition of the scope.

Example: risk of patient's safety in the operating room.

The second step is to identify potential risks that can affect the quality and the efficiency of the operating room process. In this step, we may encourage creativity and involvement of the operating room team.

The third step is the risk modeling. It consists in the development of the Bayesian networks graph (definition and choice of the variables to represent the nodes, describe the states of each node, and build the structure of Bayesian networks in terms of links between the predefined nodes) and establishment of the quantitative relation between nodes through conditional probability. In this step, we can use the hospital data source and the expert's judgment to feed the model.

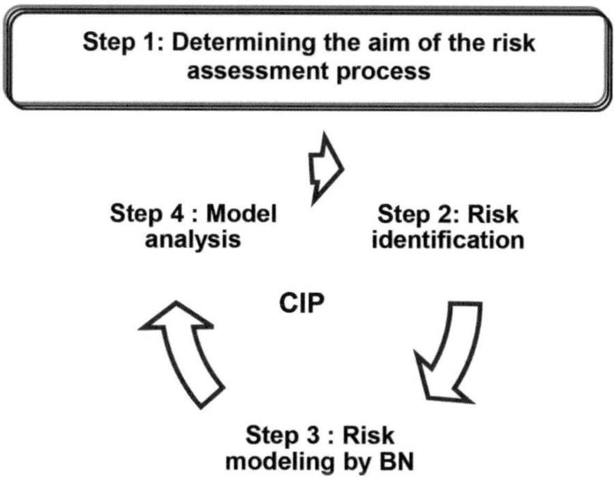

Figure 1. Methodology of risk analysis for operating room using Bayesian network.

The last step is the analysis of the results: The model should give the best understanding of the risk problem. It is useful to discuss the goodness or appropriateness of the model. It is important to validate and calibrate the model using all available source of information (expert judgment, observation, statistical data…). We should then analyze and interpret the result of risk measures to support decision-making for safety improvement.

Finally, continuous improvement efforts must incorporate a risk assessment process to ensure the effectiveness and the quality of the process. The model must be updated with the new risks and factors.

2.2. Application: patient safety risk analysis in the operating room

2.2.1. Determining the aim of the risk assessment process

The operative processes include the preoperative, intraoperative, and postoperative stages of a surgery. We are going to study the operating room processes and in particular, the intraoperative stage. It starts when the patient enters the operating room and all members of the surgical team are expected to be in the operating room at this particular time. The process ends when the patient is able to leave the operating room. During this process, the patient is monitored, anesthetized, and prepped and the operation is performed. Because of the lack of availability of actual data risk, we will forward a risk analysis based on different sources accidents described in the international literature. We will limit our study to events that cause a significant deviation of the operating room process compared to normal process and which have serious consequences for the patient (re-intervention, hospitalization in intensive care, extension of the period of hospitalization, additional care, death…).

2.2.2. Development of the Bayesian network model

To create and validate the structure of the network, we use Hugin software and more precisely Hugin Lite Evaluation.

Figure 2 illustrates the Bayesian network model of patient's safety showing interrelationships of events that may lead to patient's injury. The model has 13 nodes with one utility node. The nodes are assessed using a literature source. We present below the description of each nodes.

Surgery infection: the incidence of surgical site infections (SSI) depends upon the patient risk factors, surgical procedure, and practices observed by the operating team.

Surgical foreign body: leaving things inside the patient's body, after surgery, is an uncommon but a dangerous error. Sponges and scissors used during surgery have been left inside patients' bodies.

Operating on the wrong part of the body or wrong-site or wrong-patient or wrong-procedure surgeries: the frequency of surgery admissions experiencing a wrong site or wrong side or wrong patient or wrong procedure or wrong implant is 0.028 per 1000 admissions [6].

Medication error: wrong-dose, wrong-time, wrong-medication, or transcription errors. "A medication error is any preventable event that may cause or lead to inappropriate medication use or patient harm while the medication is in the control of the health care professional, patient or consumer. Such events may be related to professional practice, health care products, procedures and systems including prescribing, order communication, product labelling, packaging and nomenclature; compounding; dispensing; distribution; administration; education; monitoring; use" [7]. In a review of medical records from hospitals in two American states, there was a significantly higher incidence of preventable drug-related adverse events in patients aged >64 than in patients aged 16–64 years (5% compared with 3%) [8]. Errors are also significantly more likely in children.

Anesthesia equipment failure: anesthesia equipment problems may contribute to morbidity and mortality. The frequency of anesthetic equipment problems is 0.05% during regional anesthesia, and 0.23% during general anesthesia [9].

Operation error: an error may occur in surgery due to different adverse events.

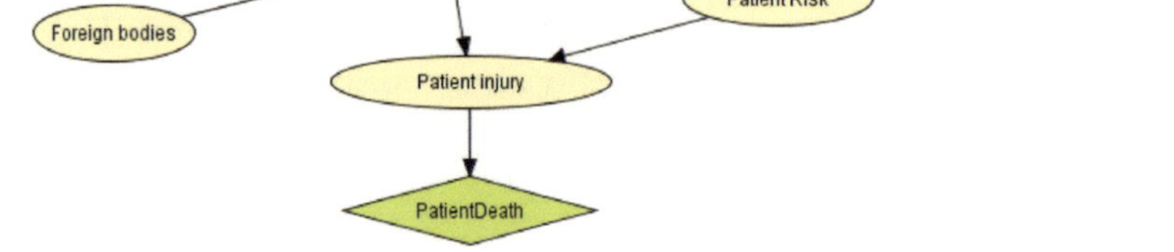

Figure 2. Bayesian network for patient safety model for the operating room.

Patient injury: an error may or may not cause an adverse event. Adverse events are injuries that cause harm to the patient (death, life-threatening illness, disability at the time of discharge, prolongation of the hospital stay, etc.).

In the following, some risk factors are given:

Patient risk: we consider two states for patient's risk, high and normal. The risk in surgery can come from patients themselves.

Age: for the age factor, we assume that the patient may be child, elderly, or adult. The age can increase the patient's risk, the risk of fall, and the risk of medication error. These risks are much higher for elderly and child than adult.

Anesthesia type: we consider two categories of anesthesia, regional and general. We assumed that "failure in anesthesia equipment" depends on anesthesia type as explained in [9].

The conditional probabilities of states of different nodes and the marginal probabilities of some adverse events have been given as input data. Each risk of adverse events is considered with two states (true if the risk exists and false if not). The probabilities are given in **Tables 1–7**.

To aggregate the impact of injuries into a single risk measure, we use utility node "Patient Death." So the task is to find the probability of patient's death after a surgery by using only the correlations and the marginal frequencies.

Operation error	**True**		**False**	
Patient risk	**High**	**Normal**	**High**	**Normal**
No	0.01	0.01	0.99	1
Small	0.18	0.81	0.009	0
Severe	0.81	0.18	0.001	0

Table 1. Conditional probability for patient injury.

Age	**Adult**	**Elderly**	**Child**
True	1.16E-5	1.16E-4	1.16E-4
False	0.9999884	0.999884	0.999884

Table 2. Conditional probability for patient fall.

Age	**Adult**	**Elderly**	**Child**
True	0.03	0.05	0.06
False	0.97	0.95	0.94

Table 3. Conditional probability for medication error.

Physical state	Weak			Normal		
Age	Adult	Elderly	Child	Adult	Elderly	Child
High	0.6	0.8	0.9	0	0	0
Normal	0.4	0.2	0.1	1	1	1

Table 4. Conditional probability for patient risk.

Anesthesia type	Regional	General
True	5×10^{-5}	2.3×10^{-3}
False	$1–(5 \times 10^{-5})$	0.9977

Table 5. Conditional probability for failure in anesthesia equipment.

Risk	Probabilities
Surgery infection	2.5×10^{-2}
Wrong site	2.6×10^{-5}
Foreign bodies	10^{-3}

Table 6. Probability of some adverse events.

Factor	State	Occurrence
Anesthesia type	Regional General	0.5 0.5
Physical state	Weak Normal	0.1 0.9
Age	Adult Elderly Child	0.5 0.2 0.3

Table 7. Probability of some factors.

2.2.3. *Analysis of the result*

After the structure of the Bayesian network is completed and probabilities are determined, the inference can be performed to estimate the probability of patient's safety risk. We conduct the calculation using Hugin software. The dependency and the correlation among risks and factors are captured in nodes "Operation error" and "Patient injury." Hence, the task is to find the probabilities of patient's death after surgery by using only the correlations and the probabilities of adverse events and the frequency of influencing factors. The probability of the death of patient is 6.37×10^{-3}. If the state of one or more variables is known, the model can be updated and the probability of patient injury and operation error will change. This should result in

Risk	Probabilities
Death of patient	6.37×10^{-3}
Death of weak physical state patient	0.02
Death of normal physical state patient	5.03×10^{-3}
Death of child patient	8.2×10^{-3}
Death of adult patient	4.98×10^{-3}
Death of elderly patient	7.08×10^{-3}

Table 8. Probability of the death of patient.

decision of not to operate the patient or postpone the surgery. For instance, the risk is much higher when the patient has a weak physical state; it is 0.02 instead of 5.03×10^{-3} for the risk of death if the patient has a normal physical state. Knowing the age of patient, we can estimate the risk of death; it is 4.98×10^{-3} for adult, 7.08×10^{-3} for elderly patient, and 8.2×10^{-3} for child (**Table 8**). It should be noted that the model and data used in this chapter have limitations. The model should be enhanced by taking into account different causes of adverse events. Data should be prevented from an adverse event database reporting system and from expert's judgment.

Several actions can be done to reduce risk and improve the safety of the patient in operating room. For instance, we can reduce the risk of retained foreign body during operation by using an appropriate sponge count and obtaining X-rays if needed to check for any retained foreign body. If we reduce this risk by 95%, the risk of the death of patient becomes 6.28×10^{-3}. Furthermore, if we reduce the risk of surgery infection by 80%, the risk of the death of patient passes to 4.5×10^{-3} instead of 6.28×10^{-3}. By acting only on "retained foreign body" and "surgery infection" adverse events, the risk can be reduced by 30%.

3. Case of the lack of data about risk

Due to the lack of data about adverse event and the fact that the adverse event reporting system does not exist, the input data of risk modeling will be provided by expert's opinion. The quality of such data must be discussed. We must help experts to provide reliable quantitative data. This can be done with the fuzzy set theory. Including the expert's judgment in the risk model is essential for providing a reliable risk picture supporting the decision-making. The second approach uses the FBN to analyze risk. Fuzzy Bayesian networks are a powerful approach for risk modeling and analysis. This is especially noticed when quantitative data are lacking and only qualitative or vague statements can be made as well when historical adverse events data are unavailable or insufficient to be used for safety assessment [10]. In this part, we present a real case of the children hospital in Rabat. To feed the model by the probabilities, we interviewed experts of the operating room. The calculation of probabilities is done out of Hugin software to conduct the fuzzy inference.

3.1. Methodology of risk analysis for the operating room using fuzzy Bayesian network (FBN)

In the following, a methodology of risk analysis of the operating room using FBN is proposed. The methodology follows five steps (**Figure 3**) and is part of the continuous improvement process (CIP). The first three steps are the same as the first proposed methodology explained above.

The fourth step is the fuzzy assessment of probability. We investigate the expert's judgment to feed the model. Experts use a linguistic variable to describe the probabilities of occurrence of adverse events. We transform the linguistic expressions into fuzzy numbers. Since we have more than one expert, we must aggregate the different opinions. For that, we use the weight of the expert to take into account the reliability of the data.

The last step is the analysis of the results: we should then analyze and interpret the results of risk measures to support decision-making for safety improvements.

Finally, the model must be implemented in Upgrading way as explained in the first method.

3.2. Application: patient safety risk analysis in the operating room

3.2.1. Risk modeling

Let us consider the previous example that we modify according to expert's opinion. **Figure 4** illustrates the BN model of patient's safety after modification. It shows interrelationships of events that may lead to patient's injury. The model has eight nodes with one utility node added to estimate the risk of the patient's death after surgery due to an error.

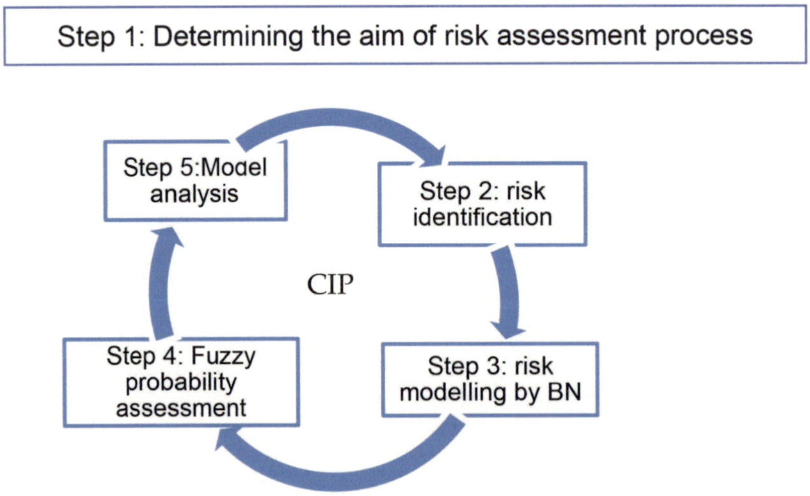

Figure 3. Methodology of risk analysis for operating room using fuzzy Bayesian network.

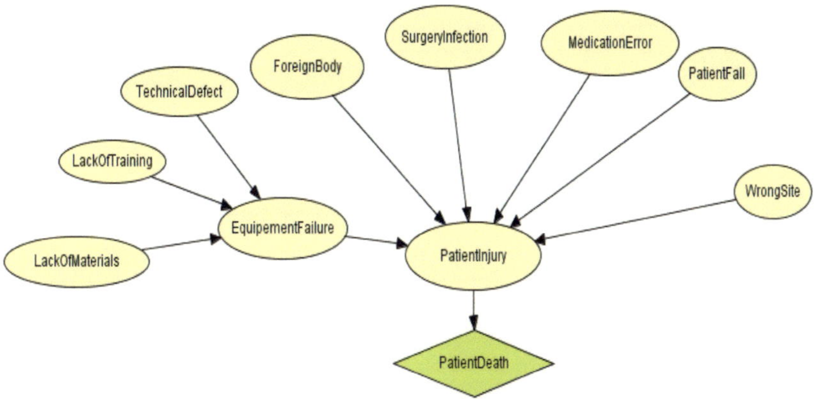

Figure 4. Bayesian network for patient safety model for the operating room.

3.2.2. Fuzzy probability assessment

Surgeons and operating team of the children's hospital IBN SINA of RABAT Morocco were asked to give judgments about the fuzzy probabilities regarding all the nodes. They use linguistic terms to describe the fuzzy probabilities and then refine them with membership functions. For example, "Very low" was assigned to node "PatientFall" and "Average" was assigned to technical defect and then were defined by the membership function (a, b, c). The other probabilities are given in **Table 11** according to the answers given by experts. The likelihood of each criterion (**Table 9**) was represented by a range of five discrete values identified by the following linguistic terms: "extremely low" (L1), "very low" (L2), "low" (L3), "average" (L4), and "high" (L5). The severity of each adverse event (**Table 10**) was represented by a range of five discrete values identified by the following linguistic terms: "negligible" (S1), "minor" (S2), "medium" (S3), "major" (S4), and "catastrophic" (S5). These five values represent the states of the node "patient's injury."

We interviewed three individuals from the operative team (surgeon, crew chief, and anesthesia nurse). They have a different point of view and confidence level toward their own subjective judgments due to the difference in background, working experience, and risk attitudes. Thus, a certain deviation exists in the data reliability among different interviewed individuals.

Set	Linguistic variable	Meaning
L1	Extremely low	Never seen
L2	Very low	One time in my career
L3	Low	Occur in another hospital
L4	Average	Occur in our hospital
L5	High	Occur in my domain

Table 9. Scale of the likelihood.

Set	Linguistic variable	Meaning
S1	Negligible	Consequence minor without prejudice (simple delay)
S2	Minor	Incident with prejudice (disorganization)
S3	Medium	Incident with impact postponement, prolongation of hospitalization, not expected transfer in reanimation)
S4	Major	Serious Consequence (re-intervention; permanent or partial disability)
S5	Catastrophic	Very serious Consequence (disability, death)

Table 10. Scale of the severity.

Expert	Weight
E1	W1 = 1/2
E2	W2 = 1/3
E3	W3 = 1/6

Table 11. Weight of expert's opinion.

Table 11 represents the weight of each expert. Expert 1 has more experience and more precise answers about adverse events than the others, so he was given the higher weight 1/2, 1/3 was assigned to expert 2, and 1/6 to expert 3.

To deal with the deviation of experts answers, the aggregated fuzzy importance of each criterion, whose properties are used to produce a scalar measure of consensus degree, is computed by the weight of the criteria according to the judgment of the expert (Eq. (1)).

$$M_1 = \begin{pmatrix} \mu^{e1b1} & \cdots & \mu^{ekb1} \\ \vdots & \ddots & \vdots \\ \mu^{e2bn} & \cdots & \mu^{ekbn} \end{pmatrix} \quad (1)$$

The expert's judgment about the likelihood and the severity of adverse events is given in **Table 12**. For instance, the probability ("high," "L5") and the severity ("catastrophic," S5) have been assigned to the node "foreign body" by expert E1; expert E2 had a different judgment about the likelihood of the same event (L3, "Low"). As you can see, experts have different opinions; that is why we used the weight of each expert.

Table 13 represents the fuzzification of the probabilities linguistic variable. For example, the triangular fuzzy number $(0.00, 10^{-8}, 2 \times 10^{-8})$ is assigned to the linguistic variable ("Extremely low," "L1"). The point $(10^{-8}, 1)$,with membership grade of 1, is the mean value; 0 and 2×10^{-8} are the left hand and right hand spreads of the triangular number, respectively (**Table 13**).

M_2 represents the vector of probabilities of basic nodes obtained using Eq. (2) and the matrix of fuzzy probabilities estimated by experts and the weight of each expert are given in **Table 5**. This step aims to determine the fuzzy probabilities of basic events.

$$M_2 = \begin{pmatrix} \mu^{b1}(x) \\ \vdots \\ \mu^{bn}(x) \end{pmatrix} = \begin{pmatrix} \mu^{e1b1} & \cdots & \mu^{ekb1} \\ \vdots & \ddots & \vdots \\ \mu^{e2bn} & \cdots & \mu^{ekbn} \end{pmatrix} \times \begin{pmatrix} w1 \\ \vdots \\ wk \end{pmatrix} \quad (2)$$

Table 14 describes the conditional probability of the node "Equipment Failure" represented by the variable N1, this variable has two states, namely true if the risk exists and false if not. If one of the three events B1, B2, and B3 occurs, the risk exists. 1f and 0f represent the crisp values 1 and 0 considered here as fuzzy number 1f (1,1,1) and 0f (0,0,0).

Nodes	Variable	E1		E2		E3	
		L	S	L	S	L	S
Lack of training	B1	L4	S3	L3	S3	L4	S4
Lack of materiel	B2	L4	S3	L3	S3	L4	S4
Technical defect	B3	L4	S3	L3	S2	L4	S4
Patient fall	B4	L2	S3	L3	S2	L1	S2
Medication error	B5	L5	S5	L3	S5	L2	S3
Surgery infection	B6	L5	S4	L3	S4	L3	S3
Foreign body	B7	L5	S5	L3	S5	L2	S4
Wrong site	B8	L4	S4	L3	S4	L2	S3

Table 12. Expert's judgment about the likelihood and the severity of adverse events.

Set	Linguistic term	Function
L1	Extremely low	$\mu_1(x) = (0.00, 10^{-8}, 2 \times 10^{-8})$
L2	Very low	$\mu_2(x) = (1.5 \times 10^{-8}, 10^{-7}, 10^{-6})$
L3	Low	$\mu_3(x) = (0.9 \times 10^{-6}, 10^{-5}, 2 \times 10^{-5})$
L4	Average	$\mu_4(x) = (1.5 \times 10^{-5}, 10^{-4}, 2 \times 10^{-4})$
L5	Very high	$\mu_5(x) = (1.5 \times 10^{-4}, 10^{-3}, 2 \times 10^{-3})$

Table 13. Fuzzification of likelihood.

N1	B4	B5	B6	B7	B8	S1	S2	S3	S4	S5
True	False	False	False	False	False	0_f	0_f	1_f	0_f	0_f
False	True	False	False	False	False	0_f	1_f	0_f	0_f	0_f
False	False	True	False	False	False	0_f	0_f	0_f	0_f	1_f
False	False	False	True	False	False	0_f	0_f	0_f	1_f	0_f
False	False	False	False	True	False	0_f	0_f	0_f	0_f	1_f
False	False	False	False	False	True	0_f	0_f	0_f	1_f	0_f

Table 14. Conditional occurrence probability of "patient injury".

B1	B2	B3	N1 = True	N1 = False
True	True	True	1_f	0_f
		False	1_f	0_f
	False	True	1_f	0_f
		False	1_f	0_f
False	True	True	1_f	0_f
		False	1_f	0_f
	False	True	1_f	0_f
		False	0_f	1_f

Table 15. Conditional occurrence probability of "equipment failure".

Table 15 represents the conditional probability of the node "Patient injury," the node has five states S1–S5 according to the severity of the harm caused to the patient. Here, the conditional probability is considered as crisp value according to the expert's opinion. Based on the harm observed, experts gave a precise answer about severity.

3.2.3. *Result and sensitive analysis*

After the structure of the BN is developed and probabilities are determined, the inference can be performed to estimate the probability of patient's safety risk. The dependency and the correlation among risks and factors are captured in node "Patient injury." Hence, the task is to find the probabilities of patient's death after surgery by using the correlations and the fuzzy probabilities of adverse events. Using the fuzzy Bayesian rule, the probability that the injury severity will be catastrophic can be calculated as given in Eq. (3):

$$P(T = S5) = \sum_i P(B = b_i) \otimes P(T = S5/B = b_i) \tag{3}$$

The probability that the injury severity will be catastrophic (S5) is $(1.5 \times 10^{-4}, 10^{-3}, 2 \times 10^{-3})$. Assuming that 80% of patients having a catastrophic injury die, the probability of the death of a patient after surgery due to an adverse event is $(1.2 \times 10^{-4}, 0.8 \times 10^{-3}, 1.6 \times 10^{-3})$. Using the center of the gravity method (Eq. (4)), we obtained COG = $(8.4 \times 10^{-4}, 1/3)$. The probability of the death of a patient after surgery is the x-axis 8.4×10^{-4}.

$$Z_{COG} = \frac{\int_z \mu_A(z) z dz}{\int_z \mu_A(z) dz} \tag{4}$$

Several actions can be done to reduce risk and improve the safety of the patient in operating room. Using this model, if we reduce the risk of retained foreign body by 60%, the risk of the death of patient becomes 3.36×10^{-4}.

If the state of one or more variables is known, the model can be updated and the probability of patient injury will change.

One of the main advantages of BN is their ability to help us to conduct inverse interference. For example, it is interesting to know, when a death is observed, what the posterior probability of a patient's infection is. In addition, if the model contains more details witch integrate the main causes of adverse events, we can obtain more interesting results such as the probability of the death of the patient due to human error or lack of training or malfunction in the organization.

The model presented must be updated when new information is available to better estimate the risk of patient safety in the operating room. The model should be enhanced by taking into account different causes of adverse events. The use of adverse event database reporting system may be very useful for getting statistics and determining the probabilities of occurrence of some adverse events. The model allows integrating a mixture source of information (probabilities from database and expert's opinion).

4. Conclusion

Safety is very essential in the healthcare system. Therefore, we should use effective and flexible methods for risk analysis to improve safety. Bayesian Networks methods are used to model and analyze risk in the operating room. The second method uses, in addition to Bayesian Network, the fuzzy logic. It allows us to use the data provided by expert and deal with the vagueness and imprecision of information. Fuzzy Bayesian network seems more flexible and interpretable than conventional Bayesian network, especially in the context of lack of data concerning risk events. This approach supports human cognition using linguistic variables which is closer to reality.

The application of the two approaches has been explained by the use of a simple model. The aim of this chapter is to propose flexible and effective methods in different context (data availability and lack of data) using Bayesian network.

However, when the size of the graph is important, the model becomes incomprehensive. We can resolve that by using object-oriented Bayesian network (OOBN). OOBN is a type of Bayesian network, comprising both instance node and usual node. An instance node is a subnetwork representing another Bayesian network. Using OOBNs, a large complex Bayesian network can be constructed as a hierarchy of sub-networks with desired levels of abstract and different levels of detail [11]. For instance, we can transform the node 'surgery infection' to a sub-network by analyzing and modeling the causes of this kind of injuries. Therefore, model construction is facilitated and communication between the model's subnetworks is more effectively performed. OOBN has a better model readability which facilitates the extension and improvement of the model.

Remedy actions are always conducted by doctors and nurses upon hazardous occurrences. Timely rescue can largely reduce the practical risks of patient's injury. By contrast, delayed remedies are of less use. It is therefore necessary to take into account the time. Consideration and incorporation of time-dependent in the risk assessment to represent equipment failure or human reliability are very important. This can be done through dynamic Bayesian network (DBN) models. DBN is an extension of Bayesian network; it is used to describe how variables influence each other over time based on the model derived from past data. A DBN can be thought as a Markov chain model with many states or a discrete time approximation of a differential equation with time steps. A dynamic Bayesian network methodology has been developed to model domino effects in [12]. Another application of DBN is presented in [13] to evaluate stochastic deterioration models.

The Bayesian network presented is a model for assessing risk of patient's safety in operating room. The model aims to capture and measure risks in the background knowledge (namely common causes and observed adverse events). Including the expert's judgment in the risk model is essential for providing a reliable risk picture supporting the decision-making. The use of adverse event database reporting system may be very useful for getting statistics and determine the probabilities of occurrence of the adverse events.

Author details

Bouchra Zoullouti*, Mustapha Amghar and Sbiti Nawal

*Address all correspondence to: bouchra.zoullouti@gmail.com

Mohammadia School of Engineers, Mohammed V University of Rabat, Rabat, Morocco

References

[1] James JT. A new, evidence-based estimate of patient harms associated with hospital care. Journal of Patient Safety. 2013;**9**:122-128

[2] Kohn LT, Corrigan JM, Donaldson MS. To Err is Human: Building a Safer Health Care System. Washington, D.C.: National Academy Press; 2000

[3] Khakzad N, Khan F, Amyotte P. Safety analysis in process facilities: Comparison of fault tree and Bayesian network approaches. Reliability Engineering and System Safety. 2011; **96**:925-932

[4] Weber P, Medina-Oliva G, Simon C, Iung B. Overview on Bayesian networks applications for dependability, risk analysis and maintenance areas. Engineering Applications of Artificial Intelligence, Special Section: Dependable System Modelling and Analysis. 2012;**25**: 671-682

[5] Zoullouti B, Amghar M, Sbiti N. Risk analysis of operating room using the Bayesian network model. International Journal of Applied Engineering Research. 2015;**10**(17): 37428-37433

[6] Availbale from: http://www.ascquality.org/qualityreport.cfm#Fall

[7] NCCMERP, National Coordinating Council for Medication Error Reporting and Prevention. About Medication Errors: What is a Medication Error?. 2012. http://www.nccmerp.org/aboutMedErrors.html

[8] Thomas EJ, Brennan TA. Incidence and types of preventable adverse events in elderly patients: Population based review of medical records. British Medical Journal. 2000; **320**(7237):741-744

[9] Fasting S, Gisvold SE. Equipment problems during anaesthesia—Are they a quality problem? British Journal of Anaesthesia. 2002;**89**(6):825-831

[10] Zoullouti B, Amghar M, Sbiti N. Risk analysis of operating room using the fuzzy Bayesian network model. International Journal of Engineering, Transactions A: Basics. 2017;**30**(1): 66-74

[11] Kjærulff UB, Madsen A. Bayesian Networks and Influence Diagrams: A Guide to Construction and Analysis. New York: Springer Verlag; 2008

[12] Khakzad N. Application of dynamic Bayesian network to risk analysis of domino effects in chemical infrastructures. Reliability Engineering & System Safety. 2015;**138**:263-272

[13] Nordgard DE, San K. Application of Bayesian networks for risk analysis of MV air insulated switch operation. Reliability Engineering and System Safety. 2010;**95**:1358-1366

Chapter 5

Continuous Learning of the Structure of Bayesian Networks: A Mapping Study

Luiz Antonio Pereira Silva,
João Batista Nunes Bezerra,
Mirko Barbosa Perkusich, Kyller Costa Gorgônio,
Hyggo Oliveira de Almeida and Angelo Perkusich

Additional information is available at the end of the chapter

http://dx.doi.org/10.5772/intechopen.80064

Abstract

Bayesian networks can be built based on knowledge, data, or both. Independent of the source of information used to build the model, inaccuracies might occur or the application domain might change. Therefore, there is a need to continuously improve the model during its usage. As new data are collected, algorithms to continuously incorporate the updated knowledge can play an essential role in this process. In regard to the continuous learning of the Bayesian network's structure, the current solutions are based on its structural refinement or adaptation. Recent researchers aim to reduce complexity and memory usage, allowing to solve complex and large-scale practical problems. This study aims to identify and evaluate solutions for the continuous learning of the Bayesian network's structures, as well as to outline related future research directions. Our attention remains on the structures because the accurate parameters are completely useless if the structure is not representative.

Keywords: Bayesian network, structure learning, continuous learning, structural adaptation, structural refinement

1. Introduction

Bayesian networks (BNs) are probabilistic graphs used to deal with the uncertainties of a domain [1]. These graphs represent the random variables of this domain and their conditional dependencies. The use of Bayesian networks, also known as Bayesian belief networks, has several points to highlight. Among them, stands out the explicit treatment of uncertainty, the ease of estimating the state of certain variables given some evidence, as well as having support methods for decision analysis and quick responses by the user [2].

The application domains of Bayesian networks have been extensive [3]. A large number of applications are in the field of medicine [4, 5], being one of the most addressed. There are also applications in the field of forecasting [6], control [7], and modeling for human understanding [8]. In the context of software engineering, fields such as project planning [9], risk management [10], and quality management [11] are addressed. Motivated by the extensive application cited, methods to improve the construction of these graphic models have become a focus of research.

A Bayesian network is defined by a directed acyclic graph (DAG) and a set of parameters for this DAG (NPT). Therefore, in order to build a Bayesian network, the definition of both the graph and the NPT must be considered. Several researches are being carried out with the intention of assisting these definitions [12–14]. However, the solutions proposed are based, for the most part, on the batch process. This process is infeasible in some application domains. Companies, for example, are increasingly storing huge databases with knowledge about their business processes. New knowledge is acquired every time. It is virtually impossible to achieve a highly accurate description of the processes involved without new data being collected or a large amount of data being stored that cannot be analyzed at once. Therefore, the need arose for solutions that continuously incorporate the updated knowledge to prior knowledge.

Ref. [15] investigated the main continuous learning solutions proposed until the development of his study. A comparative analysis between incremental algorithms and an experiment to support this analysis were performed. However, extensions of these, as well as new studies, have since been developed. This chapter aims to describe and analyze existing solutions for continuous learning of Bayesian network structures. A systematic review of the literature is carried out, and the algorithms found are divided into two groups according to their concepts: refinement and structural adaptation. Some guidelines for future research are also described.

2. Learning Bayesian networks

In probability theory, a domain D and its uncertainties can be modeled by a set of random variables $D = \{X_1, \ldots, X_n\}$. Each random variable X_i has a set of possible values that combined make up the basis for the modeling of domain D. The occurrence of each possible combination is measured using probabilities that are specified by joint probability distribution, a key concept of probability theory.

In many domains, there is a high number of variables n, requiring the use of probabilistic graphical models for the definition of joint probability distribution. Bayesian networks (BNs) belong to the family of these models that are used to represent a domain and its uncertainties [1]. A BN is a directed acyclic graph (DAG) that encodes a joint probability distribution over a set of random variables D [16]. Formally, a network for D is defined by the pair $B = \{G, \theta\}$.

The first component, G, is a DAG whose vertices correspond to the random variables $X_1, \ldots, X_n$, and the edges represent directed dependencies between variables. The vertices are represented by circles. The edges are represented by arrows indicating the direction of the causal connection between the variables; nevertheless, the information can propagate in any direction in the graph [17].

The chain rule of probability, Eq. (1), can be rewritten as Eq. (2) based on conditional independence rule. Two sets of variables $\mathbf{D}_x$ and $\mathbf{D}_y$ are independent given $\mathbf{D}_z$ if $P(\mathbf{D}_x \mid \mathbf{D}_y, \mathbf{D}_z) = P(\mathbf{D}_x \mid \mathbf{D}_z)$ whenever $P(\mathbf{D}_y, \mathbf{D}_z) > 0$.

$$P(X_1, \ldots, X_n) = \prod_{i=1}^{n} P(X_i \mid X_1, \ldots, X_{i-1}) \quad (1)$$

$$P(X_1, \ldots, X_n) = \prod_{i=1}^{n} P(X_i \mid \mathbf{Pa}_i) \quad (2)$$

In Eq. (2), $\mathbf{Pa}_i$ denotes the set of parents of the variable X_i.

The second component, θ, represents the set of parameters that quantifies the network. This set contains a parameter $\theta_{ijk} = P(X_i = x_i^k \mid \mathbf{Pa}_i = \mathbf{pa}_i^j)$ for each possible state x_i^k of X_i, and for each configuration $\mathbf{pa}_i^j$ of $\mathbf{Pa}_i$. An example of a Bayesian network is shown in **Figure 1**.

In this case, it is desired to calculate the likelihood of a person having lung cancer given the history of cancer in their family and if this person is a smoker. The node probability tables (NPTs) of the parent nodes represent a prior knowledge of these variables. The NPT of the child node represents the likelihood of a person having cancer given each possible combination of values of the parent nodes.

The goal of the learning process of a Bayesian network is to find a network (or only its structure) that best encodes the joint probability distribution of a domain. Bayesian network learning can be stated as [18]:

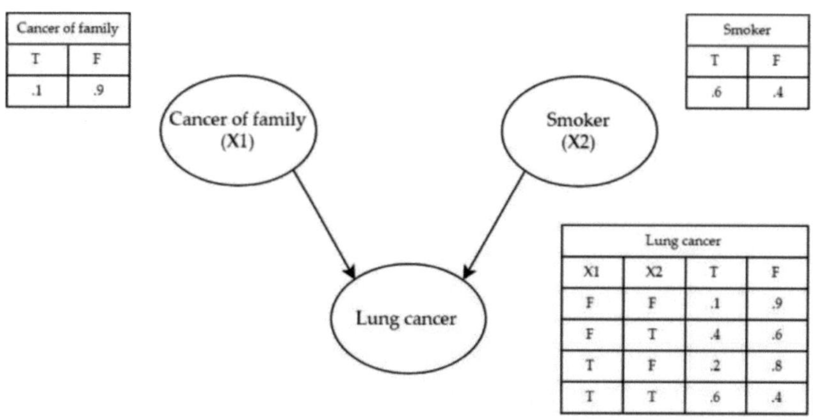

Figure 1. BN example.

Definition 1 (Bayesian network learning): *Given a data set, infer the topology for the belief network that may have generated the data set together with the corresponding uncertainty distribution.*

The learning problem of Bayesian networks can be decomposed into two subproblems: construct the structure, that is, DAG, and define the NPT [19]. Although there are many studies related to the importance of learning NPTs, this study focuses on the first subproblem described. The accurate parameters are completely useless if the structure is not representative. In [20], the importance of the structure of a network in the independence and relevance relationships between the variables concerned is described. Also, in [20], an analysis about the influence of probabilistic networks in the difficulty of representing the uncertainties present in the domain is presented.

The structure can be constructed from data only using machine learning or search techniques, such as those presented in [12, 21–23]. To optimize the definition of structure, the data can be enhanced with expert knowledge. One approach is to consult experts about the posterior probabilities of the structure to reduce the search space, such as presented in [13, 24].

In [25, 26], solutions are presented to complement a Bayesian network with the knowledge of domain experts through the addition of new factors in the model. In this way, it is possible to predict rare events, often not represented in the available databases.

Finally, the structure can be defined only according to the knowledge of specialists, where it is assumed that there are no data available before the structure construction process. In this case, the structure can be defined according to the elicited knowledge of one or multiple experts, as presented in [27, 28].

Most of the solutions to previously reported problems, as well as all of the solutions cited so far, operate as a batch process. The batch process (or batch learning) can be summarized in the delivery of a block of knowledge to an algorithm so that it learns a structure. All the knowledge available to date is used during this process. However, it is inevitable that such information will be inaccurate during the modeling of the domain [29]. For example, if knowledge is acquired from domain experts, the lack of communication between this expert and the expert on graphical models may result in errors in the Bayesian network. Similarly, if the network is being built from a data set, the data set may be inappropriate or inaccurate.

On the other hand, it is neither efficient nor, in some cases, possible to always keep the stored data in search of more representative models using batch learning algorithms [30]. To improve the use of data in the learning problem of Bayesian network structures, it's required solutions that present a continuous process of learning. In the following section, solutions for the continuous learning of the Bayesian network structures are presented.

3. Continuous learning of Bayesian networks' structure

The incentive in the application of processes that realize the learning of Bayesian networks in stages was based, initially, on the observation of the human learning by some researchers. However, the paradigm shift that provided multiple domains generated and stored more and more data also propelled its development. Continuous (or incremental) learning approaches have some widely accepted definitions found in the literature [31].

In [32], the following precise definition was stated.

Definition 2 (Incremental learner): *A learner L is incremental if L inputs one training experience at a time, does not reprocess any previous experiences, and retains only one knowledge structure in memory.*

In this definition, there are three constraints so that an algorithm can be classified as incremental. In Ref. [32], another definition with a different way of knowledge maintenance was presented.

Definition 3 (Incremental procedure): *A Bayesian network learning procedure is incremental if each iteration l, it receives a new data instance u_l and then produces the next hypothesis S_{l+1}. This estimate is then used by to perform the required task on the next instance u_{l+1}, which in turn is used to update the network and so on. The procedure might generate a new model after some number of k instances are collected.*

This definition relaxes the constraints imposed by Definition 2. For Definition 3, an incremental algorithm can be allowed to process at most k previous instances after encountering a new training instance or to keep k alternative knowledge bases in memory. In [34], another definition is based on Definition 2.

Definition 4 (Incremental algorithm): *An incremental algorithm should meet the following constraints: (i) it must require small constant time per record; (ii) it must be able to build a model using at most one scan of the data; (iii) it must use only a fixed amount of main memory, irrespective of the total number of records it has seen; (iv) it must make a usable model available at any point in time, as opposed to only when it is done with processing the data; and (v) it should produce a model that is equivalent (or nearly identical) to the one that would be obtained by the corresponding batch algorithm.*

Like previous definitions, this definition imposes constraints related to time, memory, and knowledge addressed. The Definition 4 increments the constraint related to the availability of a useful model imposed by Definition 3. Now, a lot due to its application in data streams [34], it is needed to make a usable model available at any point in time, as opposed to only when it is done with processing the data.

Based on the aforementioned definitions of an incremental learning algorithm and in [31], solutions were found that present different learning methodologies. Two groups separate these solutions. The main difference between them is in how they use the acquired knowledge. In one of these groups, denoted by refinement solutions, the data are used according to the knowledge already possessed. This knowledge is maintained in the probabilistic graphic already developed, being only refined with the new data. On the other group, denoted by structural adaptation solutions, the solutions maintain one or more candidate structures and apply to these structures the observations received. This new data set is used to update the sufficient statistics needed to build that candidate structures.

The concepts and information about the type of solutions found in this research are mapped, in an outlined (due to space constraints) and schematic way, in **Figure 2**.

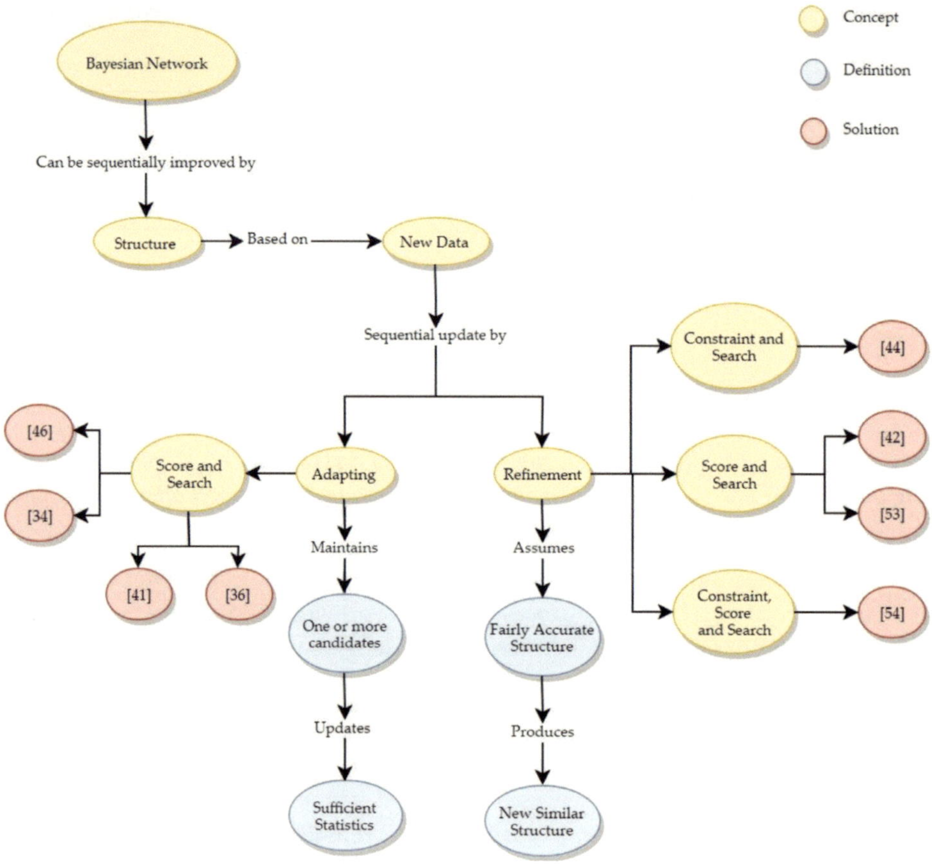

Figure 2. Mind map about solutions.

3.1. Methodology

The continuous learning Bayesian networks structure is kept like an open problem in many application domains. In this study, a systematic literature review is used to identify and evaluate solutions for the continuous learning of the Bayesian networks' structures, as well as to outline related future research directions. A combination of strings was used for title and keyword to identify articles related to continuous learning. Scopus is used as an electronic database.

In the initial search, 4150 items from Scopus were found, but only the first 400 results were checked. This stop was performed because, of these first 400 results, sorted by relevance, a sequence of 150 articles totally unrelated to the search was found. To verify this relationship, three reading steps were performed. Initially, articles were selected considering only the title and abstract. A superficial reading of the remaining articles was then performed. This step consisted of reading and interpreting section titles, figures, graphs, conclusions, and other elements. In the remaining articles, a critical reading was carried out seeking to interpret and analyze the complete text. The following sections present a description of these efforts.

3.2. Buntine's solution

In [35], a theory refinement-based approach has been proposed. The key task of theory refinement is to update the initial partial theory, usually the expert's prior domain knowledge, as far as new cases produce a posterior knowledge about the space of possible theories, being one of the fundamentals of continuous learning. Given a set of new data and total ordering of the domain variables, the solution updates both the knowledge about the structure and the parameters using different BNs.

For extending and modifying the structure, [35] proposed a batch algorithm that uses the score-and-search–based Bayesian approach. However, using some guidelines presented by the author, it is possible to convert the batch learning into continuous learning process.

The batch algorithm of [35] requires a set of ordered variables $\mathbf{X} = \{X_1, \ldots, X_n \mid X_i \prec X_{i+1}, \ldots, X_n\}$ according to the prior domain knowledge, where, for the expert, the variables that come first have influence over the others. For each variable X_i, a set of reasonable alternative parent sets $\Pi_i = \{\mathbf{Pa}_{i1}, \ldots, \mathbf{Pa}_{im}\}$ is kept according to some criteria of reasonableness. Each parent set $\mathbf{Pa}_{ij}$ is a subset of $\{Y \mid Y \prec X_i\}$.

A set of alternative parent sets Π_i for the variable X_i is denoted by the parent lattice for X_i. This parent lattice is a lattice structure where subset and superset parent sets are linked together in a web. To access all alternative parent sets $\mathbf{Pa}_j \in \Pi_i$ efficiently, only those parent sets with significant posterior probabilities are stored in the parent lattice for X_i. The root node of the parent lattice for X_i is empty set, and the leaves are the sets $\mathbf{Pa}_j$ which have no supersets contained in Π_i.

The batch algorithm also requires tree parameters $1 > C > D > E$. These are used to vary the search. The algorithm uses these parameters as base to classify the parent sets as Alive, Asleep, or Dead. The parameter C is used to separate the parent sets that finally take part on the space of alternative networks. The parameter D is used to select the reasonable alternatives to Alive parent sets. The alternatives of Alive parent sets are beams searched by the algorithm. The parameter E is used to select the reasonable alternatives to Dead parent sets. Dead parent sets are alternatives that have been explored and forever determined to be unreasonable alternatives and are not to be further explored. On the other hand, Asleep parent sets are similar but are only considered unreasonable for now and may be made alive later on.

Set the tree parameters C, D, and E to 1 will make the algorithm to be reduce to the K2 algorithm. For this, many researchers cite this algorithm as a generalization of the K2 algorithm [30, 31, 36]. At the end, a structure of alternative networks results from the set of parent sets and the network parameters, denoted by a combined Bayesian network. A pseudo-code for batch algorithm is described in [35].

To convert this batch into continuous learning algorithm, [35] describes two situations that vary according to the time available for the update. In the case where there is a short amount of time for updating the BNs, the algorithm only updates the posterior probabilities of the parent lattices. To this, it is necessary to store posterior probabilities and the counters N_{ijk} for each alternative set of Alive parent sets.

On the other hand, both structure and posterior updates are updated according to new data. For each variable X_i of the combined Bayesian network, it is necessary to: (i) update the posterior probabilities of all alive sets of the lattice, (ii) calculate the new best-posterior, and (iii) expand nodes from the open-list and continue with the search.

The generation of different networks needs to update the posterior probabilities of all alive sets of the lattice. This solution uses sufficient statistics of data that only contains counts of different entries in data instead of data entries, requiring constant time to update sufficient statistics only when new records arrive. Furthermore, [33, 35] performs an additional search over the space of alternative Bayesian networks.

3.3. Friedman and Goldschmidt's solution

Like the previous solution, [34] also addressed the problem of sequential update of the prior domain knowledge. Through the use of sufficient statistics maintained in memory for each network structure at a defined frontier, the knowledge is continuously learned. In this way, this solution provides a method that trades off between accuracy, that is, quality of structure, and storage, that is amount of information about the past observations.

In Ref. [34], three different solutions to sequentially learn BNs have been proposed. Among them, there are two extremes. The naive approach, as it is called, stores all the previously seen data and repeatedly invokes a batch learning procedure after each new observation is recorded. However, despite using as much information as possible, thus increasing the quality of the structure generated, this approach has a high storage cost. In addition, reusing batch learning increases the amount of time and processing spent.

On the other hand, the maximum a posteriori (MAP) probability approach uses a model to store all the information that is considered useful for the next steps in the knowledge update. However, the use of a single model can strongly bias the continuous learning of the model and lose information.

Aware of the disadvantages of previous approaches, [34] presents a new approach, called incremental, which proposes a tradeoff between extremes. The incremental approach does not store all data, unlike the naive approach, and it does not use a single network to represent the prior knowledge, unlike the MAP probability approach. Moreover, it allows flexible choices in the tradeoff between space and quality of the induced networks.

The basic component of this procedure is a module that maintains a set S of sufficient statistics records. The set of sufficient statistics for G, denoted by $\mathit{Suff}(G)$, can be founded by $\mathit{Suff}(G) = \{N_{X_i, Pa_i} : 1 \leq i \leq n\}$. Similarly, given a set S of sufficient statistics records, the set of network structures, denoted by $\mathit{Nets}(S)$, can be evaluated using the records in S by $\mathit{Nets}(S) = \{G : \mathit{Suff}(G) \subseteq S\}$.

http://dx.doi.org/10.5772/intechopen.80064

Two structures can easily keep track by maintaining a slightly larger set of statistics, for example, suppose on the deliberating choice between two structures G and G'. To evaluate G, in order to use a scoring function, it is needed to maintain the set $Suff(G)$. On the other hand, to evaluate G', it is needed to maintain the set $Suff(G')$. Now, supposing that G and G' differ only by one arc from X_i to X_j, note a large overlap between $Suff(G)$ and $Suff(G')$. Namely, $(G) \cup Suff(G') = Suff(G) \cup \{N_{X_j, Pa_j}\}$, where Pa_j is the parent set of Y in G'. That argument can be useful when one considers the use of the greedy hill climbing search procedure, for example. Note that it is possible to evaluate the set of neighbors of S by maintaining a bounded set of sufficient statistics.

Generalizing this discussion, the incremental approach can be applied to any search procedure that can define a search frontier. This frontier, denoted by F, consists of all the networks it compares in the next iteration. The choice of F determines which sufficient statistics are maintained in memory. After a new instance is received (or, in general, after some number of new instances are received), the procedure uses the sufficient statistics in S to evaluate and select the best scoring network in the frontier F or in $Nets(S)$. A pseudo-code for incremental approach is described in [34].

When this approach is instantiated with the greedy hill climbing procedure, the frontier F consists of all the neighbors of B_n. With bean search, on the other hand, the frontier F consists of all j candidates.

Many scoring functions can be used to evaluate the "fitness" of networks with respect to the training data and then to search for the best network. However, the incremental approach collects different sufficient statistics in different moments of the learning process. Thus, they need to compare Bayesian networks with respect to different data sets. This problem happens because, unlike [35, 34] may consider those structures that, previously, were considered as non-promising (the ones that were out of the frontier).

The two main scoring functions commonly used to learn Bayesian networks, Bayesian scores [37] and minimal description length (MDL) [38], are inappropriate for this problem. In order to overcome this problem, [34] proposed an averaged MDL measure $S'_{MDL}(G \mid D) = S_{MDL}(G \mid D)/N$, where N is the number of instances of the data set. This score measures the average encoding length per instance.

Analyzing both [35, 34] like hill-climbing searchers, they perform, for each node, operations to increase the score of the resulting structure, without introducing a cycle into the network and based on the assumption that they start from an arc-less network, as can be observed in their pseudo-code. Both stop when performing a single operation cannot increase the network's score. The difference between [35, 34] is the neighborhood composition. While [35] uses only the addition operator to construct neighbors, [34] uses the addition, reversion, and deletion of an arc.

3.4. Roure's solution

The conversions of the batch learning approaches present in [35, 34] in continuous learning approaches paved the way for popular batch algorithms like the B, K2 [37] and HCMC [39] algorithms to be turned into incremental ones [31].

In Ref. [40], two heuristics to change a batch hill-climbing search (HCS) into an incremental hill-climbing search algorithm based on combining of sufficient statistics with reduced search space have been proposed. In the batch version of HCS algorithm, a search on the space called neighborhood is performed to examine all possible local changes that can be made in order to maximize the scoring function.

Similar to the frontier presented in [34], a neighborhood of a model B consists of all models that can be build using one or more operators of a set of operators $\boldsymbol{OP} = \{$"*Add Edge*", "*Delete Edge*", "*Reverse Edge*"$\}$ and argument pairs A. Taking that into account, the sequence of operators and argument pairs added to obtain the final model B_f can be denoted by search path. Let B_0 be an initial model, a final model obtained by a hill-climbing search algorithm can be described by $B_f = op_n(\dots(op_1, A_1), \dots), A_n)$, where the search path $O_{op} = \{(op_1, A_1), \dots, (op_n, A_n)\}$ was used to build B_f.

The heuristics presented in [40] are based on two main problems: when and which part to update, and how to calculate and store sufficient statistics. The first heuristic is called by traversal operators in correct order (TOCO). TOCO verifies the already learned model and its search path for new data. If the new data alter the search path, then it is worth to update an already learned model. The second heuristic is called by reduced search space (RSS). RSS identified when the current structure needs to be revised. At each step of the search path, it stores top k models in a set B having the score close to the best one. The set B reduces the search space by avoiding to explore those parts of the space where low-quality models were found during former search steps. A pseudo-code for incremental hill-climbing search algorithm is described in [40].

3.5. Lam and Bacchus's solution

In [41], another continuous learning solution based on an extension of batch solution is presented. The batch solution used as base is presented in [42]; however, it will not be presented here because the new solution is not coupled to their batch algorithm. The proposed extension aims to perform a review of the BN structure incrementally as new data about a subset of variables were available.

This revision is done using the structure of the BN as prior probability under the implicit assumption that the existent network is already a fairly accurate model of the database. This assumption is a way to incorporate domain knowledge into the problem; however, the new refined network structure should be similar to the existing one, skewing the process.

The solution of [41] also proved, like [35] and based on the MDL measure, that if partial network structure of the whole structure gets, by changing its topology, better score to scoring function, then the whole network structure be improved if no cycles are introduced. Based on this, [41] developed an algorithm to update the BN by improving parts of it. This algorithm produces a new partial network structure based on new data set and the existing network using an extension of MDL. It then locally modifies the old structure comparing and changing correspond part according to new partial network.

The source data to algorithm consists of two components: the new data and the existent network structure. Considering the MDL principle states, finding a partial network G_p is a must that minimizes the sum of the of the length of the encoding of: (i) the partial network G_p, (ii) the new data given the network G_p, and (iii) the existent network given the network G_p. To calculate the encoding length of the first two items, there are guidelines in [41]. To calculate the length of the encoding of the third item, it is needed to compute the description of the complete existent network G given the network G_p, that is, to describe the differences between G and G_p. These differences are described by: (i) a listing of reversed arcs, (ii) the additional arcs of G, and (iii) the missing arcs of G.

A simple way to encode an arc is to describe the source node and the destination node. $2 \log n$ bits are required to describe an arc, since is required $\log n$ to identify one, provided that exists n nodes. Let r, a, and m be, respectively, the number of reversed, additional, and missing arcs in G with respect G_p, the description length G given the network G_p is $(r + a + m)\, 2 \log n$.

In order to learn the local structure, the batch algorithm proposed by [42] or other algorithm using the scoring function for each node of the partial structure is presented by Eq. (3).

$$DL_i = |\boldsymbol{Pa}_i| + \log n + \sum_{X_j \in \boldsymbol{Pa}_i} I(X_i; X_j) + (r_i + a_i + m_i)\, 2 \log n \tag{3}$$

With the third term of the equation, [41] avoided using the sufficient statistics of the old data.

After the new partial structure is learned, the review process continues with the attempt to obtain a refined structure of lower total description length with the aid of the existent structure G and the partial structure G_p. The review problem now is reduced to choosing appropriate subgraphs, denoted by the marked subgraph [41], for which we should perform parent substitution in order to achieve a refined structure of lowest total description length.

In an attempt to avoid creating cycles during each subgraph substitution, [41] uses best-first search to find the set of subgraph units that yields the best reduction in description length without generating any cycles. In addition, a list $S = \{S_1, \ldots, S_n\}$ containing a ranking of all subgraphs in ascending order of the benefit gained.

3.6. Shi and Tan's solution

In [43], an efficient hybrid incremental learning algorithm is proposed. All solutions presented so far are score-and-search–based solutions. This solution consists of a polynomial-time constraint-based technique and a hill-climbing search procedure. In this way, this solution provides a hybrid algorithm that offers considerable computational complexity savings and slightly better model accuracy.

The first fragment that composes the solution is based on a constraint-based technique. The purpose of this technique is to select candidate parent set for each variable on data. For each variable X_i, a candidate parents set S_{X_i} is set up containing all the other variables at first. If variable X_j was independent from X_i conditioned on some variables set C in previous learning procedures, the algorithm reperforms the conditional independent test and remove X_j from S_{X_i} if the independence still holds.

After this, a heuristic procedure called HeuristicIND is proposed to reduce S_{X_i} further. This procedure tries to find out a variable set to separate X_i and X_j conditionally. Using the current network structure, a tree-shaped undirected skeleton is then built up using [44]. The pseudo-codes for this procedure and for the polynomial-time constraint-based technique and a hill-climbing search procedure are described in [43].

3.7. Others continuous learning solutions

In [45], an improvement in the refinement process of [34] is presented. In this study, an incremental method for learning Bayesian networks based on evolutionary computing, denoted by IEMA, is developed.

The solutions presented so far assume that a stationary stochastic process produces all knowledge, that is, the ordering of the database is inconsequential. However, in many application domains of BNs, such as financial problems [46], the processes vary according to the time, the data are non-stationary or piecewise stationary distributed, which would reduce the adequacy of the solutions already mentioned. In [47], the assumption on stationary data is relaxed and an incremental learning Bayesian network based on non-stationary data domains is developed.

In [48], the streaming data prediction is addressed. A parallel and incremental solution for learning of BNs from massive, distributed, and dynamically changing data by extending the classical scoring and search algorithm and using MapReduce is presented.

Ref. [37] also presents an algorithm for stream and online data, more precisely, data that are privately and horizontally shared among two or more parties. This algorithm is based on an efficient version of sufficient statistics to learning privacy-preserving Bayesian networks.

In [49], an active and dynamic method of diagnosis of crop diseases has been proposed based on Bayesian networks and incremental learning. To incremental learning, a new algorithm for dynamically updating the Bayesian network-based diagnosis model over time also is proposed.

Ref. [50] transformed the local structure identification part of Max-min Hill-climbing (MMHC) algorithm into an incremental fashion by using heuristics and applied incremental hill-climbing to learn a set of candidates-parent-children for a target variable.

In [51], an incremental algorithm for BN structure learning that can deal with high dimensional domains has been proposed.

In [52], the concept of influence degree is used to describe the influence of new data on the existing BN. A scoring-based algorithm for revising a BN iteratively by hill-climbing search for reversing, adding, or deleting edges has been proposed.

In [53], an approach to incremental structure optimize is presented. Based on a specific method, this approach decomposes the initial network into several subnets created from a junction tree developed using information about the joint probability of the network. With some adaptations, it can be used as a continuous learning algorithm.

It is also important to point out that studies are found that denominate their algorithm as incremental, but these algorithms result in incremental learning of variables or small parts of the network as they are available and do not necessarily generate a new model, as it is constrained in the definition.

3.8. Comments and future research

Some comments comparing some solutions have already been made during your descriptions. However, to facilitate the understanding of the methodologies used, some techniques and characteristics were compared below. A summary table among some solutions is presented in **Table 1**.

Some features of the solutions were discussed in **Table 1**. These features are important for the differentiation of the proposal of each algorithm. It is noteworthy that, among the outstanding solutions, none proposes to use the domain specialist as a source of posterior knowledge, only as a source of prior knowledge. This knowledge can be increased over time and can be used to improve the built network. In addition, changes that are not able to be identified only with the use of data, such as adding factors and incomprehensibility of the model, can be identified.

When checking the stochastic process, it is noted that only [47] adopts a non-stationary domain, even considering the increasing diversity of domain. The other quoted solutions kept their focus on stationary domains, more common at the time of their developments. Still considering the domain, the effectiveness of the algorithms is rarely validated in a real domain, even if it is done with experiments that use data coming from real domains.

The local search is a standard present among the search procedures used. Despite their high computational complexity, methods were developed so that this was not a constraint and

Solution	Domain expert	Type of structure changes	Scoring function	Search procedure	CI	Stochastic process
[36]	Prior knowledge	Adding edges	Any	Any local search	—	Stationary
[34]	Prior knowledge	Adding, deleting, or reversing edges	Averaged MDL	Any local search	—	Stationary
[40]	Prior knowledge	Adding, deleting, or reversing edges	Any	Hill-climbing search	—	Stationary
[41]	Prior knowledge	Adding, deleting, or reversing edges	Extension of MDL	Best-first search	—	Stationary
[43]	Prior knowledge	Adding, deleting, or reversing edges	—	Hill-climbing search	InfoChi	Stationary
[47]	Prior knowledge	Adding, deleting, or reversing edges	—	Any local search	—	Non-stationary

Table 1. Summary table among some solutions.

they then continued to be used. Only one solution made use of conditional independence (CI) tests. Ref. [43] developed a new technique that is used as the basis for CI tests. Considering scoring functions, some solutions leave open the use of any function. However, an adaptation to their applications is necessary in sufficient statistics used in some, for example, in [34]. On the other hand, some solutions use the MDL measure, but already adapted, either to reduce computational complexity or to achieve better results in different data set data. Other features can be addressed in future reviews, such as computational complexity, procedure focus, and the type of application domain, among others.

4. Conclusions

This chapter presents a survey based on a systematic literature review of continuous learning solutions of Bayesian network structures. It searches articles with an algorithm or approach considered as incremental according to the definitions presented in the text.

The solutions found can be classified as structural adaptation or refinement. The first group, in short, uses the new data set to maintain sufficient statistics that are used to store the existing knowledge about the domain. The second group uses this new set of data to perform a refinement in the network, based on previous knowledge coupled in the old structure.

Finally, the presence of the posterior knowledge of the domain specialist during the incremental learning process and experiments in real domains for the validation of some solutions findings are in the future works.

Acknowledgements

The authors would like to thank Federal University of Campina Grande in Brazil for supporting this study.

Conflict of interest

The author declares that there is no conflict of interest regarding the publication of this chapter.

Author details

Luiz Antonio Pereira Silva, João Batista Nunes Bezerra, Mirko Barbosa Perkusich*,
Kyller Costa Gorgônio, Hyggo Oliveira de Almeida and Angelo Perkusich

*Address all correspondence to: mperkusich@gmail.com

Federal University of Campina Grande, Paraíba, Brazil

References

[1] Ben-Gal I. Bayesian Networks. Encyclopedia of Statistics in Quality and Reliability. John Wiley and Sons; 2008

[2] Lee E, Park Y, Shin JG. Large engineering project risk management using a Bayesian belief network. Expert Systems with Applications. 2009;**36**(3):5880-5887

[3] Daly R, Shen Q, Aitken S. Learning Bayesian networks: Approaches and issues. The Knowledge Engineering Review. 2011;**26**(2):99-157

[4] Lucas PJF, Van der Gaag LC, Abu-Hanna A. Bayesian networks in biomedicine and health-care. Artificial Intelligence in Medicine. 2004;**30**(3):201-214

[5] Shwe M, Cooper G. An empirical analysis of likelihood-weighting simulation on a large, multiply connected medical belief network. Computers and Biomedical Research. 1991;**24**(5):453-475

[6] Abramson B et al. Hailfinder: A Bayesian system for forecasting severe weather. International Journal of Forecasting. 1996;**12**(1):57-71

[7] Forbes J et al. The batmobile: Towards a Bayesian automated taxi. IJCAI. 1995;**95**:1878-1885

[8] Husmeier D. Sensitivity and specificity of inferring genetic regulatory interactions from microarray experiments with dynamic Bayesian networks. Bioinformatics. 2003;**19**(17): 2271-2282

[9] Pendharkar PC, Subramanian GH, Rodger JA. A probabilistic model for predicting software development effort. IEEE Transactions on Software Engineering. 2005;**7**:615-624

[10] Fan CF, Yu YC. BBN-based software project risk management. Journal of Systems and Software. 2004;**73**(2):193-203

[11] Jeet K, Bhatia N, Minhas RS. A Bayesian network based approach for software defects prediction. ACM SIGSOFT Software Engineering Notes. 2011;**36**(4):1-5

[12] Neapolitan RE. Learning Bayesian networks. Vol. 38. Upper Saddle River, NJ: Pearson Prentice Hall; 2004

[13] Heckerman D. A tutorial on learning with Bayesian networks. In: Learning in Graphical Models. Dordrecht: Springer; 1998. pp. 301-354

[14] O'Hagan A et al. Uncertain Judgements: Eliciting Experts' Probabilities. Chichester: John Wiley & Sons; 2006

[15] Huang H et al. A comparatively research in incremental learning of Bayesian networks. Intelligent Control and Automation. 2004. WCICA 2004. Fifth World Congress on. Vol. 5. IEEE, 2004

[16] Friedman N, Geiger D, Goldszmidt M. Bayesian network classifiers. Machine Learning. 1997;**29**(2-3):131-163

[17] Amari S. The Handbook of Brain Theory and Neural Networks. London, England: MIT Press; 2003

[18] Sangüesa R, Cortés U. Learning causal networks from data: A survey and a new algorithm for recovering possibilistic causal networks. AI Communications. 1997;**10**(1):31-61

[19] Neil M, Fenton N, Nielson L. Building large-scale Bayesian networks. The Knowledge Engineering Review. 2000;**15**(3):257-284

[20] Druzdel MJ, Van Der Gaag LC. Building probabilistic networks: Where do the numbers come from? IEEE Transactions on Knowledge and Data Engineering. 2000;**12**(4):481-486

[21] van Dijk S, Van Der Gaag LC, Thierens D. A skeleton-based approach to learning Bayesian networks from data. European Conference on Principles of Data Mining and Knowledge Discovery; Berlin, Heidelberg: Springer; 2003

[22] Tsamardinos I, Brown LE, Aliferis CF. The max-min hill-climbing Bayesian network structure learning algorithm. Machine Learning. 2006;**65**(1):31-78

[23] Zhang Y, Zhang W, Xie Y. Improved heuristic equivalent search algorithm based on maximal information coefficient for Bayesian network structure learning. Neurocomputing. 2013;**117**:186-195

[24] Heckerman D, Geiger D, Chickering DM. Learning Bayesian networks: The combination of knowledge and statistical data. Machine Learning. 1995;**20**(3):197-243

[25] Constantinou AC, Fenton N, Neil M. Integrating expert knowledge with data in Bayesian networks: Preserving data-driven expectations when the expert variables remain unobserved. Expert Systems with Applications. 2016;**56**:197-208

[26] Constantinou A, Fenton N. Towards smart-data: Improving predictive accuracy in long-term football team performance. Knowledge-Based Systems. 2017;**124**:93-104

[27] Hu X-X, Wang H, Shuo W. Using expert's knowledge to build Bayesian networks. Computational Intelligence and Security Workshops, 2007. CISW 2007. International Conference on. IEEE; 2007

[28] Richardson M, Domingos P. Learning with knowledge from multiple experts. In: Proceedings of the 20th International Conference on Machine Learning (ICML-03); 2003

[29] Lam W. Bayesian network refinement via machine learning approach. IEEE Transactions on Pattern Analysis & Machine Intelligence. 1998;**3**:240-251

[30] Zeng Y, Xiang Y, Pacekajus S. Refinement of Bayesian network structures upon new data. International Journal of Granular Computing, Rough Sets and Intelligent Systems. 2009;**1**(2):203-220

[31] Alcobe JR. Incremental methods for Bayesian network structure learning. AI Communications. 2005;**18**(1):61-62

[32] Langley P. Order effects in incremental learning. In: Learning in Humans and Machines: Towards an Interdisciplinary Learning Science. Vol. 136. Pergamon; 1995. p. 137

[33] Friedman N, Goldszmidt M. Sequential update of Bayesian network structure. In: Proceedings of the Thirteenth conference on Uncertainty in artificial intelligence; Morgan Kaufmann Publishers Inc.; 1997

[34] Domingos PM, Hulten G. Catching up with the data: Research issues in mining data streams. DMKD. 2001

[35] Buntine W. Theory refinement on Bayesian networks. In: Proceedings of the Seventh conference on Uncertainty in Artificial Intelligence; Morgan Kaufmann Publishers Inc.; 1991

[36] Samet S, Miri A, Granger E. Incremental learning of privacy-preserving Bayesian networks. Applied Soft Computing. 2013;**13**(8):3657-3667

[37] Cooper GF, Herskovits E. A Bayesian method for the induction of probabilistic networks from data. Machine Learning. 1992;**9**(4):309-347

[38] Lam W, Bacchus F. Learning Bayesian belief networks: An approach based on the MDL principle. Computational Intelligence. 1994;**10**(3):269-293

[39] Kočka T, Castelo R. Improved learning of Bayesian networks. In: Proceedings of the Seventeenth conference on Uncertainty in artificial intelligence; Morgan Kaufmann Publishers Inc.; 2001

[40] Alcobé JR. Incremental hill-climbing search applied to Bayesian network structure learning. In: Proceedings of the 15th European Conference on Machine Learning; Pisa, Italy; 2004

[41] Lam W, Bacchus F. Using new data to refine a Bayesian network. Uncertainty Proceedings. 1994;**1994**:383-390

[42] Castelo R, Kocka T. On inclusion-driven learning of Bayesian networks. Journal of Machine Learning Research. 2003;**4**:527-574

[43] Shi D, Tan S. Incremental learning Bayesian network structures efficiently. In: Control Automation Robotics & Vision (ICARCV), 2010 11th International Conference on. IEEE; 2010

[44] Chow CK, Liu CN. Approximating discrete probability distributions with dependence trees. IEEE Transactions on Information Theory. 1968;**14**:462-467

[45] Tian F et al. Incremental learning of Bayesian networks with hidden variables. In: Data Mining, 2001. ICDM 2001, Proceedings IEEE International Conference on. IEEE; 2001

[46] Shi D, Tan S. Incremental learning bayesian networks for financial data modeling. In: Intelligent Control, 2007. ISIC 2007. IEEE 22nd International Symposium on. IEEE; 2007

[47] Nielsen SH, Nielsen TD. Adapting Bayes network structures to non-stationary domains. International Journal of Approximate Reasoning. 2008;**49**(2):379-397

[48] Yue K et al. A parallel and incremental approach for data-intensive learning of Bayesian networks. IEEE Transactions on Cybernetics. 2015;**45**(12):2890-2904

[49] Zhu Y et al. Mathematical modelling for active and dynamic diagnosis of crop diseases based on Bayesian networks and incremental learning. Mathematical and Computer Modelling. 2013;**58**(3-4):514-523

[50] Yasin A, Leray P. iMMPC: a local search approach for incremental Bayesian network structure learning. International Symposium on Intelligent Data Analysis; Berlin, Heidelberg: Springer; 2011

[51] Yasin A, Leray P. Incremental Bayesian network structure learning in high dimensional domains. In: Modeling, Simulation and Applied Optimization (ICMSAO), 2013 5th International Conference on. IEEE; 2013

[52] Liu W et al. A Bayesian network-based approach for incremental learning of uncertain knowledge. International Journal of Uncertainty, Fuzziness and Knowledge-Based Systems. 2018;**26**:87-108

[53] Chunsheng G, Qiquan S. Incremental structure optimize of Bayesian network based on the lossless decomposition. In: Artificial Intelligence and Computational Intelligence (AICI), 2010 International Conference on. Vol. 2. IEEE; 2010

Chapter 6

Multimodal Bayesian Network for Artificial Perception

Diego R. Faria, Cristiano Premebida, Luis J. Manso,
Eduardo P. Ribeiro and Pedro Núñez

Additional information is available at the end of the chapter

http://dx.doi.org/10.5772/intechopen.81111

Abstract

In order to make machines perceive their external environment coherently, multiple sources of sensory information derived from several different modalities can be used (e.g. cameras, LIDAR, stereo, RGB-D, and radars). All these different sources of information can be efficiently merged to form a robust perception of the environment. Some of the mechanisms that underlie this merging of the sensor information are highlighted in this chapter, showing that depending on the type of information, different combination and integration strategies can be used and that prior knowledge are often required for interpreting the sensory signals efficiently. The notion that perception involves Bayesian inference is an increasingly popular position taken by a considerable number of researchers. Bayesian models have provided insights into many perceptual phenomena, showing that they are a valid approach to deal with real-world uncertainties and for robust classification, including classification in time-dependent problems. This chapter addresses the use of Bayesian networks applied to sensory perception in the following areas: mobile robotics, autonomous driving systems, advanced driver assistance systems, sensor fusion for object detection, and EEG-based mental states classification.

Keywords: Bayesian networks, machine learning, multimodal robotic perception

1. Introduction

Bayesian networks (BNs) allow a tractable graph-based representation for probabilistic reasoning (or inference), under uncertainty, about a given problem or domain. A recurrent problem in robotics is to reason about the class of an object in the environment, given evidence (from sensors, e.g. RGB-D cameras), and probabilistic models (e.g. probability outputs of a classifier) in the domain that represents the problem. For example, a robot would be needed to detect and then recognise a particular type of object (such as a mug) in a given place (e.g. kitchen) [1]. Another example would be an autonomous vehicle that has to detect road users; hence, the object categories of interest would be pedestrian, cyclist, car and van, bus and truck, and motorised two-wheelers.

The topology, or structure, of a BN graph is the first step in solving the problem, and it should provide the relationship (dependencies represented by links) among the nodes (variables in the problem domain). The next step is to define the conditional probabilities for the nodes and, then, the joint probability of the BN has to be considered in order to allow computing the posterior probability of the form $P(\text{Class} \mid \text{Evidences})$, i.e. *a posteriori* of the class, or category, given the evidences from a set of sensor-based models [1].

In this chapter, we will address BN with similar topologies to the one illustrated in **Figure 1**. The structure shown in **Figure 1** is a 'common effect' chain [2], which means that all parent nodes contribute to the node C designated by the 'class'. The node C is the label variable and takes values such as: $C = \{person, non\text{-}person\}$, or $C = \{1, 0\}$, or in multiple class case, $C = \{mug, spoon, knife, fork, plate, can\}$ or $C = \{concentrated, relaxed, neutral\}$. The evidence nodes, as illustrated in **Figure 1**, provide probability values per class of interest; thus, such nodes are modelled by a classifier (e.g. convolutional neural network [CNN], SVM, and Bayes classifier). The node called 'context' might represent evidence from the environment, or information shared by the infrastructure (e.g. cameras mounted on the scenario), or any other evidence not directly related to a given learning classifier using data/features from sensors onboard the robot.

The remainder of this chapter is organised as follows: Section 2 briefly describes the use of BNs for supervised classification problems. Use cases on object manipulation, pedestrian classification, and EEG-based Mental State Classification are described in Sections 3–5, respectively. Finally, Section 6 presents a summary and remarks.

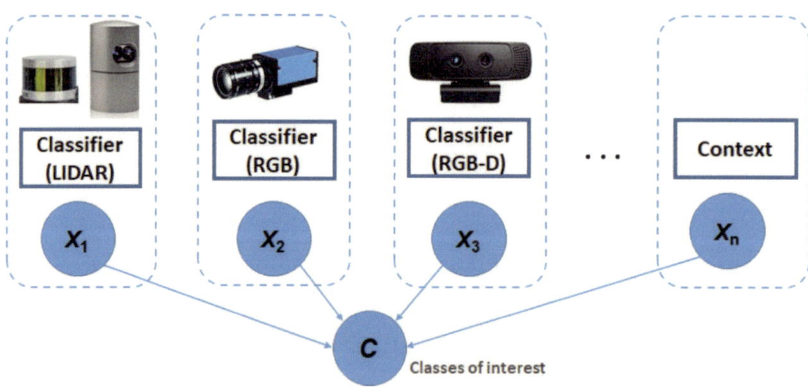

Figure 1. Topology of a BN where all the parent nodes $\{X_1, X_2, \ldots, X_n\}$ contribute to a common effect, which is the set of classes (node C) of interest in a given robotic domain.

2. Bayesian networks for supervised classification

In a more general and high-level perspective, a BN is characterised by **nodes** that represent a finite set of random variables, i.e. a variable/function whose outputs outcome from a random measured process (belonging to the domain of interest) and **links** (i.e. directed arcs) that represent the direct dependencies between the nodes. Hereafter, the link dependencies will assume the form of conditional probabilities. By examining **Figure 1**, we can see that each node X_i, $i = 1, \ldots, n$, is conditionally independent of all the nodes, while the node C is conditionally dependent of all its parent nodes $X_1, X_2, \ldots, X_n$. This is a simple BN structure where the nodes represent learning model—in the form of supervised classifiers. On the other hand, which is common in some classification problems, the nodes may represent features as extracted from observed data.

Let $P(X \mid C)$ be the outcome of a learning classifier given the observed/measured sensor data. The variable X can represent a learned model, using a probabilistic classifier, based on supervised and measured data from a camera, a LIDAR, an RGB-D sensor, or a combination of multimodality data. We could represent the conditional probability as $P(X, \theta \mid C)$ to explicitly show that the output also depends on a learning model (here represented by θ).

In a nutshell and considering the use cases described in the sequel, BNs are used to express the joint probability of events (represented by the nodes) that model a classification system where the relationships between events are expressed by conditional probabilities. Given the observations/measurements (evidence) and prior knowledge, statistical inference is accomplished using the Bayes' theorem. The goal is to calculate the posterior $P(C \mid X_i)$ of the set of classes given the evidential nodes X_i, $i = 1, \ldots, n$. So, inference in supervised classification applications aims to estimate the probability of the classes given the class-conditional probabilities, priors and observations.

In this work, in one of the approaches presented, we consider BN structures where the sensory data are transformed to a feature space which is then feed into a trained classifier. The classifier is assumed to output a class-conditional probability which is then used to calculate the *a posteriori*. When multiple sensors are considered, the conditional independency property between sensors will be satisfied, for example: $P(X_{sensor1} \mid X_{sensor2}, X_{sensor3}) = P(X_{sensor1})$. The following sections will present three different study cases where BN are employed as a supervised classifier.

3. Use case in object shape representation through in-hand exploration

Accurate modeling of the world (environment and its components) is important in autonomous robotics applications. More precisely, for grasping applications dealing with objects used in everyday tasks, the object information (intrinsic and extrinsic) acquired before the robot executes a task is crucial for grasp strategies. The object geometry (size and shape) plays an important role in such applications, where its representation is also valuable for classification into a class of known objects and also for identification of regions on the object surface proper for a stable grasp. Since the robotic end-effector usually relies on the knowledge of object geometry to plan or to estimate grasp candidates, the more accurate the geometry of the object, the higher is the likelihood of success when estimating the candidate's grasp for that object. Many techniques can be used to reconstruct and represent an object using different sensors, such as vision-based systems, laser range finders, etc., where the most common is through visual information.

Mapping techniques such as occupancy grid [3, 4] have been used in robotics to describe the environment of mobile robots. Two-dimensional grids have been used for static indoor mapping as shown in [5]. The idea is to estimate the probability of each cell to be occupied or empty after the sensors' observation. Probabilistic volumetric maps are also useful in robotics by providing means of integrating different occupancy belief maps in order to update a central multimodal map using Bayesian filtering. A grid divides the workspace into equally sized voxels, and the edges are aligned with one of the axes of a reference coordinate frame. The coverage of each voxel given the sequence of batches of measurements is modelled through a probability density function. The probabilistic approach for building volumetric maps of unknown environments can also be based on information theory. Each sensor (e.g. vision, laser, etc.) can adopt an entropy gradient-based exploration strategy to define the occupied regions (most explored) in the map.

Object in-hand exploration is the procedure of exploring the shape of objects using tactile information and fingers motion around the object surface to reconstruct its shape [6]. In order to acquire the probabilistic representation of an object using a volumetric map, it is necessary to have an *a priori* estimation of the area, where the object is placed for mapping. There are two scenarios in which in-hand exploration can be applied: a static object placed at a specific location or an object being explored in-hand in constant motion (dynamic exploration with moving object). The sensors used for this task is a cyberglove that measures fingers flexure (0–255 range), with six electromagnetic motion sensors (Polhemus sensors), where each sensor provides 6D information (x, y, z, *yaw*, *pitch*, and *roll*), and tactile sensors in each fingertip and palm (Tekscan pressure sensor) that measure the force (0–255 range). **Figure 2** depicts the experimental setup and sensors used for object in-hand exploration.

When the object exploration is in-hand and the object is moving, then it is needed to perform a registration to map the object displacements into a single frame of reference. We can consider that, for every motion of the object, a local map is built, so that all local maps should be integrated into a global map to have the whole representation of the object shape exploration in the same frame of reference. Knowing the object initial position and the object displacements, we can compute the transformations to have all points in the same frame of reference. Given that the sensor attached to the object has six DoF (x, y, z, *yaw*, *pitch*, and *roll*), we can compute the rotation and translation of the object. We compute the rotation matrix of the object in a specific point in time using α = yaw (rotation in z axis), β = pitch (rotation in y), and φ = roll (rotation in x).

To map the point cloud in the same frame of reference, for all points, we find the translation of the fingertip sensor to the object sensor and then we apply the rotation to that point, $p' = R_o t$, where p' is the new position of the 3D point that we are mapping to the same frame of reference of the object sensor; R_o is the rotation matrix 3×3 of the object sensor; and t the translation of the fingertip sensor to the object sensor.

http://dx.doi.org/10.5772/intechopen.81111

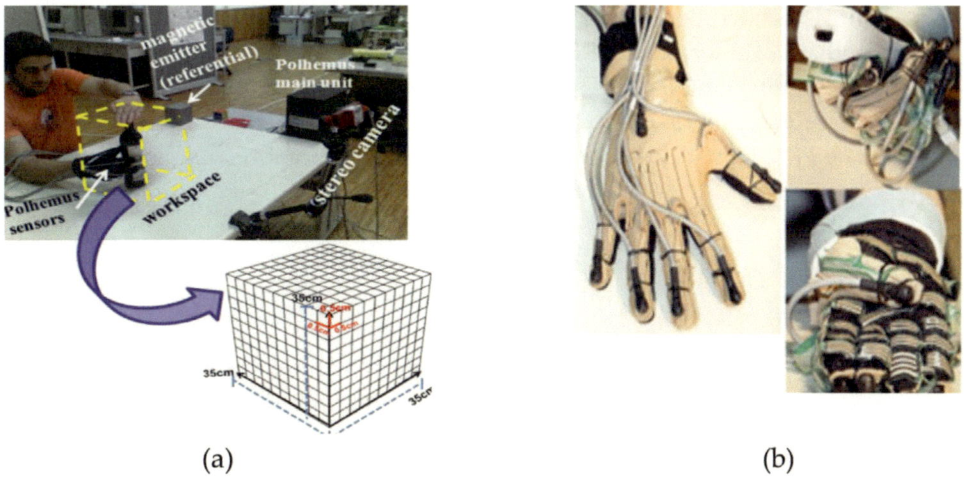

Figure 2. Experimental setup: (a) workspace for mapping (grid 35 cm × 35 cm × 35 cm equally divided, where each voxel is sized with 0.5 cm); (b) Polhemus Liberty Motion Tracking System: magnetic sensors attached to the cyberglove (fingertips and back of the hand).

The Bayesian volumetric map [6] is an occupancy grid, i.e. discrete random fields, wherein each cell has an assigned value, which represents the probability of the cell being occupied. The dimensions of the voxels define the spatial resolution of the representation. The edges of the grid are aligned with one of the axes of the world frame of reference *W*. In this work, the map is a 3D grid comprised of a set of cells $c \in M$, denoted as voxels, wherein each voxel is a cube with edge $\varepsilon \in R$. The voxels divide the workspace into equally sized cubes with volume ε^3. The occupancy of each individual voxel is assumed to be independent from the other voxels occupancy, and thus, *Oc* is a set of independent random variables as follows:

- $c \in M$: Index a cell on the Map;
- $Oc \in [0, 1]$: Probability describing if the cell c is empty or occupied;
- *Zc*: Measurement that influences the cell *c*. It represents the measurements acquired from five sensors, each one returns the 3D location of each finger movement in the map;
- P(*Oc*): Probability distribution of preliminary knowledge describing the occupancy of the cell c, initially as a uniform distribution (0.5 for each state: empty or occupied); and
- P(*Zc*|*Oc*): Probability density function corresponding to the set of measurements that influences the cell c taken from the in-hand exploration measurements. This distribution is computed from the in-hand exploration sensor model.

The knowledge about the occupancy of a voxel *c* in the map *M*, after *Z* measurements received at time *t* from the sensors, is represented by the probability density function $P(Oc|Z_c^t)$. Updating the 3D probabilistic representation of the manipulated object shape upon a new measurement Z^t means updating the probability distribution function $P([Oc = 1]|Z_c^t)$ of the voxel *c* influenced by the measurement *Z* at time *t*. Voxels are influenced by a measurement

Z^t if the location associated with the sample computed from the sensor model P (Z_c^t|[Oc = 1]) is contained in that voxel location c. For each voxel c, the set of measurements Z_c^t contains n measurements Z_c influencing a voxel c along the time t. The probability density function of the object shape representation of voxel c given the Z_c measurements influencing such voxel is represented by $P(Z_c^t \mid [Oc = 1])$. To update the occupancy estimation of a cell in the map, the Bayes rule is applied:

$$P([Oc = 1] \mid Z_c^t) = \frac{P(Z_c^t \mid [Oc = 1])P([Oc = 1])}{P(Z_c^t \mid [Oc = 0])P([Oc = 0) + P(Z_c^t \mid [Oc = 1])P([Oc = 1])}, \tag{1}$$

where $P([Oc = 0]) = 1 - P([Oc = 1])$; $P(z_c^t \mid [Oc = 1])$ is given by the probability density function computed from the sensor model and $P(z_c^t \mid [Oc = 0])$ is a uniform distribution.

Assuming that consecutive measurements Z^t are independent given the cell occupancy, the following expression is obtained:

$$P([Oc = 1] \mid Z_c^t) = \beta \times P(Oc) \prod_{t=1}^{T} P(Z_c^t \mid [Oc = 1]), \tag{2}$$

where β is a constant representing a normalization, factor ensuring that the left side of the equation sums up to one over all Oc.

The cells occupancy in the map are probabilities that are updated over time as long as the sensors measurements are active. At the end of the in-hand exploration of the object, the cells are allowed to represent only two states: occupied or empty, $Oc \in [0, 1]$, so that a threshold is used for each cell to consider one of the two states:

$$Oc = \{0, P(Z_c^t) < 0.5\ 1, P(Oc \mid Z_c^t) \geq 0.5\}. \tag{3}$$

Figure 3 shows an example of the probabilistic volumetric map and its utility. The map can be used to represent the full model of the object as well as partial volume of the object and contact.

Each magnetic sensor attached to the fingertips returns the 3D coordinates of the finger location based on the sensor frame of reference (source/emitter of the Polhemus Liberty tracking system). The frame rate of each sensor was defined to be up to 15 Hz. During data acquisition, a workspace (35 cm^3) is defined in the experimental area for mapping. The grid space is divided into equally sized voxels (also denoted as cells) of 0.5 cm^3. Due to the size of each cell, relative to the standard deviation of the magnetic tracking sensors measurements (up to 3 mm), inside each cell a 3D isotropic Gaussian probability distribution is defined, $P(z_c^t \mid Oc)$, centred at the cell central point with the standard deviation 0.3 cm and mean value equal to the central point coordinates of the cell. In other words, this means that the model attempts to ensure that, upon receiving a measurement from the sensor attached to the fingertip, the closer the finger position is to the centre of a specific cell of the map, the more probable that cell is occupied. Furthermore, during the object surface exploration, the more often that the finger passes through that cell, the cell probability is updated with higher certainty in which that given point position actually belongs to the object surface. The probability that a measurement belongs to a cell is given by a normal distribution using the known sensor position error as the standard deviation and the sensors positions relative to the centre of each cell in the map as follows:

$$P(Oc) = \frac{1}{2\,\Pi^{3/2}\,|\Sigma|^{1/2}} e\left(-\frac{1}{2}(x-\mu)^T \Sigma^{-1}(x-\mu)\right), \quad (4)$$

where $P(Z_c^t | Oc)$ represents the probability distribution of the sensor measurement given a specific cell Oc; $|\Sigma|$ represents the determinant of Σ (sensor noise variation). It can also represent a scalar value. After normalization, it takes the form:

$$P([Oc = 1]) = exp\left(-\frac{(x-u_x)^2 + (x-u_y)^2 + (x-u_z)^2}{2\,\sigma^2}\right), \quad (5)$$

where (x, y, z) are the coordinates of the 3D point on the object surface, and u is the central coordinate of the cell (for each axis). The in-hand exploration of objects can be performed by using the thumb and other fingers, i.e. the occupancy grid can be influenced by them over time, thus, expanding on the model for cell update, the contribution of the sensor on each finger through time can be made explicit on the decomposition as follows:

$$P\left(Z_{thumb}^{t=0}, \ldots, Z_{thumb,}^{T} Z_i^{t=0}, \ldots, Z_i^T, Oc\right) = P(Oc)\prod_{t=0}^{T} P\left(z_{thumb}^{t} | Oc\right)\prod_{i=1}^{N} P\left(z_i | Oc\right), \quad (6)$$

where T represents the current time instant and $N = 4$, the remaining four fingers of the hand. This process for updating the cell over time recursively (i.e. initially using the cell probability as a uniform distribution: empty or occupied, and later the cell probability—updated with the Bayes rule—is used as prior for the next update), represents a Bayesian network.

The BN representation of the formalism applied to the decomposition of the joint distribution in which the sensor model was used is shown in **Figure 4**. The plate notation relies on assumptions of duplicated subgraph as many times as the associated repetition number (in

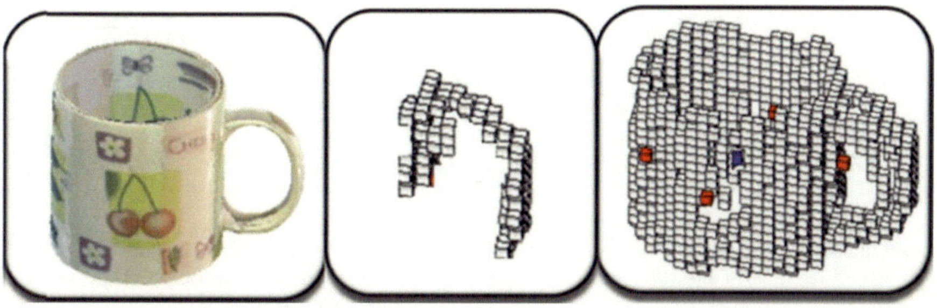

Figure 3. Examples of the Bayesian volumetric map. Left image: real object; middle image: partial volume of the object obtained during in-hand exploration; right image: map of the full object model and contact points overlaid on the object surface (red voxels representing the contact points and blue voxel representing the centroid of the object to define its frame of reference).

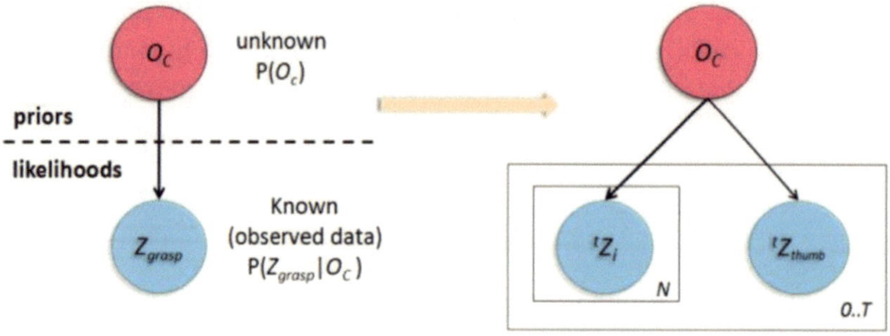

Figure 4. BN for object representation by in-hand exploration using occupancy grid. The left image shows the labels: prior, posterior, and respective distributions, yet not necessary in dynamic BN representations. The variables are defined in terms of their notation and conditional dependence. The instantiation is defined with their parameters and the random variables that support the model are fully described (i.e. their significance and measurable space). The right image shows the plate notation applied to the BN formalism to represent the in-hand exploration of objects, making explicit the contribution of the sensors over time.

Figure 5. Object representation using the probabilistic volumetric map: sponge and its computed map.

Figure 6. Object shape representation by in-hand exploration of a spray bottle. The first image (left to right) is the raw data (point cloud), next three images are different views of the voxels representation of the object shape, and the last image is the occupancy representation of the cells, the darkest ones represent the lower probabilities (less explored regions).

this particular case the hand fingers); the variables in the subgraph are indexed according to the repetition number; the links that cross a plate boundary are replicated for each subgraph repetition; the distributions are in the joint distribution as an indexed product of the sequence of variables. Bayesian formalisms for probabilistic model construction and some BN examples of occupancy grid model can also be seen in [6, 7].

Figures 5 and **6** shows different household objects explored in-hand for shape retrieval.

4. Use case in pedestrian classification

A pedestrian detection system is one of the key components in Advanced Driver Assistance Systems (ADAS) and also in autonomous driving vehicles. Recently, pedestrian detection has regained particular attention from academia, automotive industry, and society [8]. In this chapter, pedestrian classification is studied based on a multimodal Bayesian network, where the BN's structure has a node representing the binary class (pedestrian and nonpedestrian) and the parent nodes are represented by machine learning models in the form of supervised classifiers. In terms of sensory data, we will consider a LIDAR sensor as an intermodality technology, which provides range (distance) and reflectance (intensity return). In order to study multimodality between two sensor technologies, a colour (RGB) camera is also considered in the BN. The classifiers are modelled by a deep convolutional neural network (CNN). Data from a LIDAR enter into the CNN classifier in the form of high-resolution distance/depth (DM) and reflectance maps (RMs). Distance and intensity (reflectance) raw data from the LIDAR are transformed to high-resolution (dense) maps as described in [9, 10].

A multimodal BN is then used to combine the likelihoods from CNN-classifiers learned using data from a LIDAR (based on DM and RM) and from a camera. Pedestrian recognition is evaluated on a 'binary classification' dataset created from the KITTI Vision Benchmark Suite, which provides data from a colour camera and from a Velodyne HDL-64E LIDAR. The performance results using the BN are compared with the CNNs having a single modality as input, and against nonlearning rules, namely: minimum, maximum, and average.

We will formulate the classification problem in such a way that the class node (C) of the BN is inferred from the classification nodes (X_{RGB}, X_{DM}, X_{RM}); therefore, the 'full' joint distribution is expressed by:

$$P(C, X_{RGB}, X_{DM}, X_{RM}) = P(C)P(X_{RGB} \mid C)P(X_{DM} \mid X_{RGB}, C)P(X_{RM} \mid X_{DM}, X_{RGB}, C), \tag{7}$$

assuming each classifier node contributes independently to explain C and also assuming the classifiers are independent of each other but not independent of the class so, e.g. $P(X_{DM} \mid X_{RGB}, C) = P\left(X_{DM} \mid C\right)$, we can express the class-conditional a posteriori as:

$$P(C \mid X_{RGB}, X_{DM}, X_{RM}) \sim P(C)P(X_{RGB} \mid C)P(X_{DM}, C)P(X_{RM}, C). \tag{8}$$

We will consider the class a-priori probability to be uniform and equally distributed; thus, the probability of being pedestrian or nonpedestrian ($P(C)$) can be dropped out from the equation above. Therefore, the inference problem resumes to a product of the outputs probabilities from the CNN models.

To evaluate the multimodal BN described here, a pedestrian classification dataset was created based on the 2D object-detection dataset of KITTI. The labelled classes are given in the form of 2D bounding box tracklets: 'Pedestrian', 'Car', 'Truck', 'Tram', 'Van', 'Person (sitting)', 'Cyclist' and 'Misc'. The classes were separated in two categories of interest: pedestrian and nonpedestrian, i.e. a binary problem. The number of positives examples is 4487 cropped images (labelled bounding boxes of type 'Pedestrian'), while the negative class has 47,378 cropped images (types: 'Cyclist'. 'Car', 'Person (sitting)' and so on). It was considered 70% for the training set (10% of that for validation) and the remaining 30% for the testing set. **Table 1** gives a summary of the dataset used in this use case.

Among several convolutional neural networks, we opted to use AlexNet CNN architecture with batch normalization in the first two layers and the last layer, the *softmax* activation function with two classes and dropout of 50%. The network was trained from scratch for the pedestrian and nonpedestrian classes [10]. Through the bounding boxes provided by the KITTI dataset, we cropped the objects contained in the depth and reflectance maps images. All objects were resized to the size of 227 × 227 because this is the network input size. The network was trained with the following parameter settings: 30 epochs, batch size equal 64, stochastic gradient descent optimizer with $lr = 0{:}001$ (learning rate), $decay = 10 - 6$ (learning rate decay over each update), $momentum = 0.9$, and categorical cross-entropy as loss function.

Denoting $P(X_i \mid C)$ the confidence (i.e. the class-conditional probability) yielded by deep models CNN*i* ($i = 1, \ldots, n$), where n is the number of models, CNN1 and CNN2 denote CNN models learned from DM and RM (reflectance), respectively, while CNN3 denotes a model using RGB data. Three nonlearning fusion rules are considered: average (AVE), maximum (MAX), and minimum (MIN). The average rule calculates the simple mean of the CNN-classifiers outputs $F\text{-}ave = \frac{1}{n}\sum_{i=1}^{n} P(X_i \mid C)$. The maximum rule outputs the maximum value over the classifier responses, $F\text{-}max = \max\{P(X_i \mid C)\}$, while the minimum rule is $F\text{-}min = min\{P(X_i \mid C)\}$.

The pedestrian classification results are reported using Precision (Pre), Recall (Rec), and F-score (F1) performance measures, allowing a more detailed and accurate analysis of the results. The F-scores values were obtained considering a threshold of 0.5. A number of pedestrian and nonpedestrian examples are unbalanced, as shown in **Table 1**; thus, F-score is here considered because it is a suitable performance measure for unbalanced cases. The results obtained using the BN and the rules AVE, MAX, and MIN are shown in **Figure 7**.

Summary of dataset for pedestrian classification	
Training set	n# positives = 2827
	n# negatives = 29,849
Validation set	n# positives = 314
	n# negatives = 3316
Testing set	n# positives = 1346
	n# negatives = 14,213

Table 1. Pedestrian dataset.

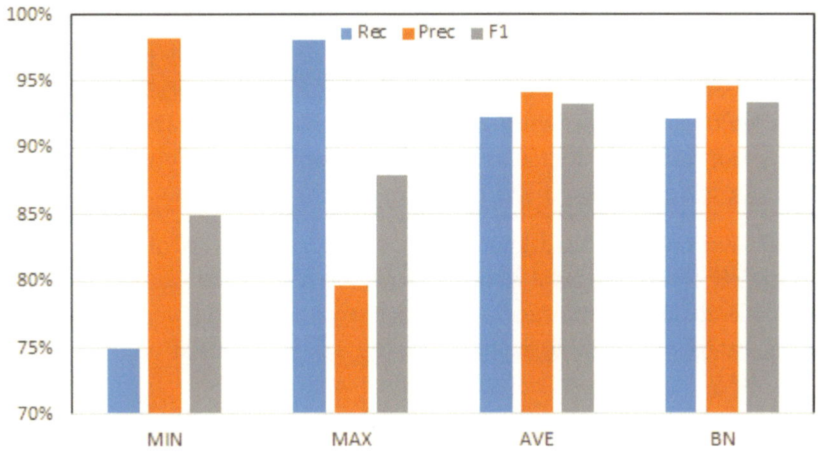

Figure 7. Classification performance, in terms of Pre, Rec, and F-1, considering a multimodal BN in comparison with deterministic rules (Min, Max, Average).

Results show that decision rules like minimum and maximum tend to have poor results, in terms of F-score, compared to the average rule and the multimodal BN. However, the values of Precision and Recall (or True Positive rate) are very high for Min and Max, respectively. The Average and the BN achieved close classification performance in all measures, although the BN's results were slightly better.

5. Use case in EEG-based mental states classification

AI-enabled wearable technology has the ability to enhance the capabilities of today's user-centred devices and analytics toward promoting humans' quality of life and enabling an improved health care by monitoring humans' complex bio-signals, reducing risks, detecting anomalous situations, thus, optimising standards of care. A good example is the EEG-based brain-controlled devices that can serve as powerful aids for severely disabled people in their daily life, especially to help them to move voluntarily. The EEG-based brain-machine interfaces are one of the many alternatives that can be used to interact with devices using the superficial brain activity signals. These signals, called electroencephalograms or EEG for short, convey information regarding the voltage measured by electrodes (dry or wet) placed around the scalp of an individual. Recently, new applications for restoring function to those with motor impairments using EEG-based brain machine interfaces for conveying messages and commands to devices such as robot arm, wheelchair, and any other devices using bio-signals have been developed. A good example where EEG is employed is to detect mental states. The ability to autonomously detect mental states, whether cognitive or affective, is useful for multiple purposes in many domains such as robotics, health care, education, neuroscience, etc. The importance of efficient human-machine interaction mechanisms increases with the number of real life scenarios where smart devices, including autonomous robots, can be applied. One of the many alternatives that can be used to interact with machines is through superficial brain activity signals. A major challenge in brain-machine interface applications is inferring how momentary mental states are mapped into a particular pattern of brain activity. One of the main issues of classifying EEG signals is the amount of data needed to properly describe the different states, since the signals are complex. The signals are considered stationary only within short intervals, that is, why the best practice is to apply short-time windowing technique in order to detect local discriminative features to meet this requirement.

This section presents how Bayesian inference can be used to classify mental states. The framework consists of (i) statistical and temporal features extraction using time window technique, (ii) attributes selection to keep only the relevant information from the signals, and (iii) Bayesian classification technique to categorise multiple mental states (e.g. relaxed, neutral, and highly concentrated).

5.1. Data acquisition

The sensor Muse Headband was used for data collection. The Muse is a commercial EEG sensing device with five dry-application sensors, one used as a reference point (NZ, at the centre of the forehead) and four (at points TP9, AF7, AF8, TP10, i.e. around the forehead **Figure 8**) to record brain wave activity. To prevent the interference of electromyographic signals, nonverbal tasks that required little to no movement were set. Blinking, though providing interference to the AF7 and AF8 sensors, was neither encouraged nor discouraged to retain a natural state. This was due to the dynamicity of blink rate being linked to tasks requiring differing levels of concentration, and as such, the classification algorithms would take these patterns of signal spikes into account. In addition, subjects were asked not to close their eyes during any of the tasks. Three stimuli were devised to cover the three mental states available from the Muse Headband—relaxed, neutral, and concentrating. A dataset was created after five participants performing the three mental states, where each session lasted 1 minute. The relaxed task had the subjects listening to low-tempo music and sound effects designed to aid in meditation while being instructed on relaxing their muscles and resting. For a neutral mental, a similar test was carried out, but with no stimulus at all, this test was carried out prior to any others to prevent lasting effects of a relaxed or concentrative mental state. Finally, for concentration, the subjects were instructed to follow the 'shell game' in which a ball was hidden under one of the three cups, which were then switched, the task was to try and follow which cup hid the ball. After a short amount of time into the stimulus starting, as to not gather data with an inaccurate class, the EEG data from the Muse Headband were automatically recorded for 60 seconds. The data were observed to be streaming at a variable frequency within the range of 150–270 Hz.

5.2. Feature extraction

Feature extraction and classification of EEG signals are primary goals in brain–computer interface (BCI) applications. One challenging problem when it comes to EEG feature extraction is the complexity of the signal. Nonstationary signals can be observed during the change in alertness and wakefulness, during eye blinking, and also during transitions of mental states. Discriminative features rely on statistical techniques such as mean, standard deviation, autocorrelation, statistical moments of third and fourth order (skewness and kurtosis

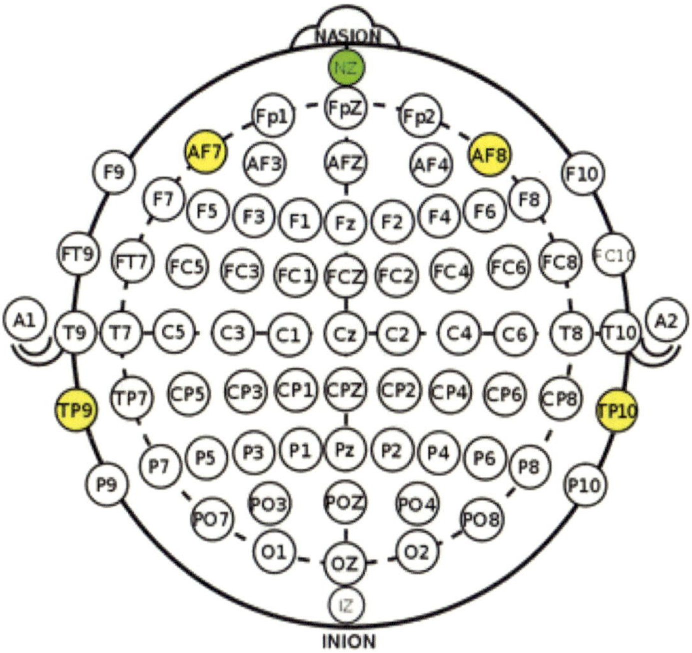

Figure 8. The International 10-20 EEG electrode placement standard [11]. The sensors of the Muse Headband are denoted in yellow. The NZ placement (green) is used as a reference for calibration.

to measure the asymmetry of the data and also the peakedness of the probability distribution of the data), time-frequency based on fast Fourier transform (FFT), Shannon entropy, max-min features in temporal sequences, log-covariance given a set of statistical data, and derivatives of the features from different time instants. These features are computed in terms of the temporal distribution of the signal in a time window of 1 second, with overlap of half second between the sliding windows. Details about the modeling and implementation of the features can be found in [12]. Another important point to compute the features is the signals from the EEG Muse headband. Since it returns five types of signal frequencies (alpha, beta, theta, delta, and gamma), then, we compute all set of features for each signal. The aforementioned set of features for all signals are around 2100 feature values. In order to reduce and optimise the classification performance, feature selection is needed.

5.3. Feature selection

There are various well-known algorithms for features selection in the state of the art. These types of algorithms aim at reducing the number of attributes present in a dataset while retaining a model's predictive accuracy. The following algorithms were used to compare the accuracy performance when used with a Naïve Bayes classifier (NB) and a Bayesian Network (BN): (i) *OneR* calculates error rate of each prediction based on one rule and selects the lowest risk classification [13]; (ii) *Information Gain* assigns a worth to each individual attribute by measuring the information gain with respect to the class (difference of entropy) [14]; and (iii) *Evolutionary*

Algorithm creates a population of attribute subsets and ranks their effectiveness with a fitness function to measure their predictive ability of the class [15]. At each generation, solutions are bred to create offspring, and weakest solutions are killed off in a tournament of fitness.

5.4. Classification

Two models were trained on Bayes' theorem, a formula of conditional probability based on hypothesis H and evidence E. The theorem states that the probability of the hypothesis being true before evidence $P(H)$ is related to the probability of the hypothesis after reading the evidence $P(H|E)$ and is given as follows: $P(H|E) = \frac{P(E|H)P(H)}{\sum_i P(E|H)P(H)}$. A simplistic Naive Bayes model has been used, which has a non-cosideration of the relationships between the features models. It uses the maximum a posteriori decision rule $\hat{y} = \text{argmax}_{k\in\{1,\ldots,K\}} P(C^k)\prod_{i=1}^{n} P(x_i|C^k)$. A BN (*Bayes Net*) model was also trained. This method generates a probabilistic graphical model via representing probabilities of variables to classes on a directed acyclic graph (DAG) as follows: $P(C^{t-1:t-T}|X^{t:t-T}) = \frac{1}{\beta}\prod_{k=t}^{T-t} P(X^k|C^k)P(C^k)$. The goal is to infer the current time value of C^t given the data $X^{t:t-T} = \{X^t, X^{t-1}, \ldots, X^{t-T}\}$ and the prior knowledge of the class, which is attained by the a-posteriori probability $P(C^t|C^{t-1:t-T}, X^{t:t-T})$. The superscript notation denotes the set of values over a time interval.

5.5. Experimental results

The five generated sets from the original dataset classified by NB and BN are shown in **Table 1**. The most effective model for this EEG dataset using Bayesian inference was the BN along with the *OneR Attribute Selector*, which had a high accuracy of 73.67% using around 2% of the total of features extracted when classifying the data into one of the three mental states. For each test, 10-fold cross-validation was used to train the model. The lowest performance is 54.2% (*Information Gain* dataset with a NB classifier). It is reasonable to assume that the naivety in not considering attribute relationships has led to poorer results. These preliminary results show that a BN can be considered for EEG data classification. However, other methods of classification can achieve better performance with the same set of features. In order to improve the performance, we can adopt the strategy of fusion of multiple classifiers using the Bayes' theorem for fusion as shown in [1, 16] **Table 2** presents the result of Bayesian inference combined with feature selection algorithms. Better results are attained when using OneR algorithm for features selection followed by classification via Bayesian networks.

Dataset	Accuracy %		
	Naive Bayes	Bayesian network	Number of selected features (%)
OneR	**56.30**	**73.67**	44 (2.05)
Information gain	54.20	71.64	31 (1.44)
Evolutionary algorithm	55.04	70.31	99 (4.61)

Table 2. Accuracy of trained models.

6. Summary

Approaches based on Bayesian network (BN) have been described considering three case studies: Bayesian volumetric map for object perception, pedestrian classification for autonomous-vehicles perception and for EEG-based mental states classification. BNs were formulated and applied in supervised pattern classification problems. In all cases, the BNs assumed conditional independence between sensors' modalities or feature models.

In summary, this chapter has addressed BN with examples, where other machine learning techniques were employed and combined with BN to sensory perception in applications related to robotics (multimodal sensor fusion for object detection), advanced driver assistance systems for autonomous driving systems, and EEG-based mental states classification, which can be used to control devices (e.g. robots) or in health-related areas for mental health monitoring.

Acknowledgements

This work has been partially supported by the MICINN Project TIN2015-65686-C5-5-R, by the Extremaduran Government project GR15120, by the FEDER project 0043-EUROAGE-4-E (Interreg V-A Portugal-Spain - POCTEP), and by Fundação Araucária (CONFAP Brazil) with a mobility grant to Dr Diego R. Faria and Professor Eduardo P. Ribeiro to coordinate the project "Steppingstones to transhumanism: merging EEG-EMG data to control a low-cost prosthetic hand".

Author details

Diego R. Faria[1]*, Cristiano Premebida[2], Luis J. Manso[1], Eduardo P. Ribeiro[3] and Pedro Núñez[4]

*Address all correspondence to: d.faria@aston.ac.uk

1 School of Engineering and Applied Science, Aston University, UK

2 Institute of Systems and Robotics, University of Coimbra, Portugal

3 Department of Electrical Engineering, Federal University of Parana, Brazil

4 School of Technology, University of Extremadura, Cáceres, Spain

References

[1] Premebida C, Faria DR, Nunes UJ. Dynamic Bayesian network for semantic place classification in mobile robotics. In: Autonomous Robots (AURO). Springer; 2017;**41**(5):1161-1172

[2] Korb KB, Nicholson AE. Bayesian Artificial Intelligence. 2nd ed. Boca Raton, FL, USA: CRC Press, Inc.; 2010

[3] Moravec HP. Sensor fusion in certainty grids for mobile robots. AI Magazine. 1988; **9**(2):61-74

[4] Elfes A. Using occupancy grids for mobile robot perception and navigation. IEEE Computer. 1989;**22**:46-57

[5] Thrun S, Burgard W, Fox D. Probabilistic robotics. In: Arkin RC, editor. MIT press; 2005. ISBN 9780262332750

[6] Faria DR. Probabilistic learning of human manipulation of objects towards autonomous robotic grasping [PhD thesis]. Portugal: Department of Electrical and Computer Engineering, University of Coimbra; 2014

[7] Faria DR, Martins R, Lobo J, Dias J. Probabilistic representation of 3D object shape by in-hand exploration. IEEE/RSJ International Conference on Intelligent Robots and Systems; 2010

[8] Ross PE. Uber robocar kills pedestrian, despite presence of safety driver. In: IEEE Spectrum. Available from: https://spectrum.ieee.org [Accessed: June 2018]

[9] Premebida C, Garrote L, Asvadi A, Pedro Ribeiro A, Nunes U. High-resolution LIDAR-based depth mapping using bilateral filter. IEEE International Conference on Intelligent Transportation Systems ITSC; 2016

[10] Melotti G, Premebida C, Goncalves N, Nunes U, Faria DR. Multimodal CNN pedestrian classification: A study on combining Lidar and camera data. 21th IEEE International Conference on Intelligent Transportation Systems (ITSC), 2018

[11] Klem GH, Lüders HO, Jasper HH, Elger C. The ten-twenty electrode system of the International Federation. The International Federation of Clinical Neurophysiology. Electroencephalography and Clinical Neurophysiology. Supplement. 1999;**52**:3-6

[12] Bird JJ, Manso LJ, Ribeiro EP, Ekart A, Faria DR. A study on mental state classification using EEG-based brain-machine interface. In: 9th IEEE International Conference on Intelligent Systems; 2018

[13] University of Waikato. OneR [online] Weka.sourceforge.net. Available from: http://weka.sourceforge.net/doc.dev/weka/classifiers/rules/OneR.html [Accessed: August 2018]

[14] University of Waikato. InfoGainAttributeEval [online] Weka.sourceforge.net. Available from: http://weka.sourceforge.net/doc.dev/weka/attributeSelection/InfoGainAttributeEval.html [Accessed August 9, 2018]

[15] Back T. Evolutionary Algorithms in Theory and Practice: Evolution Strategies, Evolutionary Programming, Genetic Algorithms. Oxford University Press; 1996

[16] Premebida C, Faria DR, de Souza FA, Nunes UJ. Applying probabilistic mixture models to semantic place classification in mobile robotics. In: IEEE International Conference on Intelligent Robots and Systems; 2015. pp. 4265-4270

Quantitative Structure-Activity Relationship Modeling and Bayesian Networks: Optimality of Naive Bayes Model

Oleg Kupervasser

Additional information is available at the end of the chapter

http://dx.doi.org/10.5772/intechopen.85976

Abstract

Previously, computational drag design was usually based on simplified laws of molecular physics, used for calculation of ligand's interaction with an active site of a protein-enzyme. However, currently, this interaction is widely estimated using some statistical properties of known ligand-protein complex properties. Such statistical properties are described by quantitative structure-activity relationships (QSAR). Bayesian networks can help us to evaluate stability of a ligand-protein complex using found statistics. Moreover, we are possible to prove optimality of Naive Bayes model that makes these evaluations simple and easy for practical realization. We prove here optimality of Naive Bayes model using as an illustration ligand-protein interaction.

Keywords: quantitative structure-activity relationship, Naive Bayes model, optimality, Bayes classifier, Bayesian networks, protein-ligand complex, computational drag design, molecular recognition and binding, ligand-active site of protein, likelihood, probability

1. Introduction

The determination within the chapter is based on a paper [1]. Bayes classifiers are broadly utilized right now for recognition, identification, and knowledge discovery. The fields of application are, for case, image processing, personalized medicine [2], chemistry (QSAR (quantitative structure-activity relationship) [3, 4]; see **Figure 1**). The especial importance Bayes Classifiers have in Medical Diagnostics and Bioinformatics. Cogent illustrations of this can be found in the work of Raymer and colleagues [5].

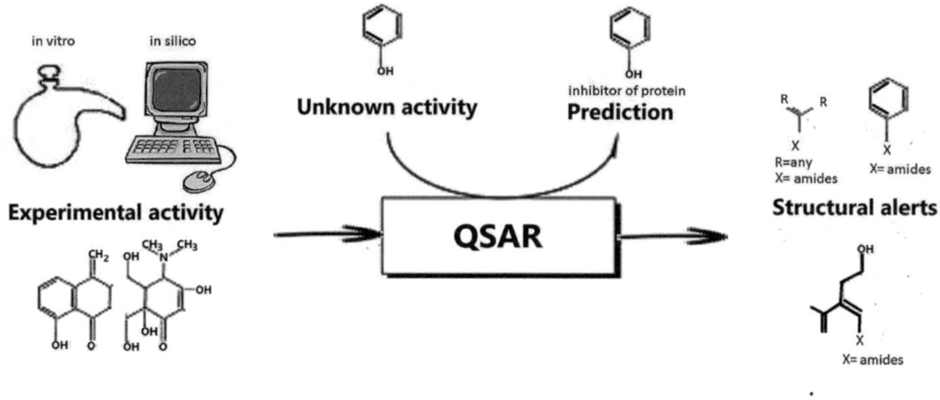

Figure 1. Quantitative structure-activity relationship.

Let us give some example of using QSAR from papers [3, 4]:

"Molecular recognition and binding performed by proteins are the background of all biochemical processes in a living cell. In particular, the usual mechanism of drug function is effective binding and inhibition of activity of a target protein. Direct modeling of molecular interactions in protein-inhibitor complexes is the basis of modern computational drug design but is an extremely complicated problem. In the current paradigm, site similarity is recognized by the existence of chemically and spatially analogous regions from binding sites. We present a novel notion of binding site local similarity based on the analysis of complete protein environments of ligand fragments. Comparison of a query protein binding site (target) against the 3D structure of another protein (analog) in complex with a ligand enables ligand fragments from the analog complex to be transferred to positions in the target site, so that the complete protein environments of the fragment and its image are similar. The revealed environments are similarity regions and the fragments transferred to the target site are considered as binding patterns. The set of such binding patterns derived from a database of analog complexes forms a cloudlike structure (fragment cloud), which is a powerful tool for computational drug design."

However, these Bayes classifiers have momentous property—by strange way the Naive Bayes classifier more often than not gives a decent and great description of recognition. More complex models of Bayes classifier cannot progress it significantly [1]. In the paper [6] creators clarify this exceptional property. In any case, they utilize a few suspicions (zero–one misfortune) which diminish all-inclusiveness and simplification of this proof. We allow in this chapter a common verification of Naive Bayes classifier optimality. The induction within the current chapter is comparative to [1]. The consequent attractive consideration of Naive Bayes classifier optimality problem was made in [7, 8]. Be that as it may, shockingly these papers do not incorporate any investigation of the past one [1].

We would like to prove Naive Bayes classifier optimality using QSAR terminology. Indeed, we use QSAR only for clearness; the proof is correct for any field of use of Naive Bayes classifier.

Let us define the essential issue that we attempt to unravel within the chapter. Assume that we have a set of states for a complex of ligand-active site of protein and a set of factors that characterize these states. For each state, we know the likelihood dispersion for each factor. In any case, we have no data of the approximate relationships of the factors. Presently, assume that we know factor values for some test of the state. What is the probability that this test corresponds to some state? It could be a commonplace issue of recognition over a condition of incomplete data.

In the simplest case, we can define two states for "ligand-active site of protein" complex. It is 0 (ligand is not bound to active site of protein) or 1 (ligand is not bound to active site of protein).

The next step is definition of factors (reliabilities below) that characterize strength of a bond for "ligand-active site of protein" complex. Let us grant an illustration of factors (reliabilities below) from experience of QSAR in papers [3, 4]:

"First, consider the protein 5 A°-environment $A = \{a_1, a_2, \ldots a_N\}$ of one ligand atom X in the analog protein, that is, all atoms from the binding site that are in the 5 A°-neighborhood of X. Suppose that the complete target binding site T consists of N′ atoms: $T = \{t_1, t_2, \ldots t_{N'}\}$ and there exists a subset $T_0 \subseteq T$ of size n ($N' \geq n \geq 4$) such that n atoms from T_0 are similar to n atoms $A_0 = \{a_{i1}, a_{i2}, \ldots a_{in}\} \subseteq A$ in their chemical types and spatial arrangement. The search for A_0 and T_0 is performed using a standard clique detection technique in the graph whose nodes represent pairs (a_i, t_i) of chemically equivalent atoms and edges reflect similarity of corresponding pairwise distances. If the search is successful, the optimal rigid motion superimposing matched protein atoms is applied both to the initial ligand atom X and its complete environment A (**Figure 2(a)** in [3]). The atoms are thus transferred to the target binding site. Then we extend the matching between A_0 and T_0 by such atom pairs (a_i,t_i) that a_i and t_i have the same chemical atom type in the coarser 10-type typification mentioned above, and the distance between t_i and the image a'_i of atom a_i is below a threshold. Next, a reliability value R, with $0 \leq R \leq 1$, is assigned to the image X′ of X in the target site and reflects the similarity between the environments of X and its image X′. If the environments are highly similar ($R \approx 1$) we expect that the position of X′ is the place where an atom with chemical type identical to X can be bound by the target, since the environment of X′ contains only atoms required for binding with no "alien" atoms. However, as illustrated in **Figure 2(a)** in [3], the analog site may contain extra binding atoms (shown on the lower side) that decrease the reliability value. In a simple form, the reliability R can be defined as the sum of the number of matched atoms divided by the total number of analog and target atoms in the 5 A°-environments of X and X′, respectively (**Figure 2(b)** in [3]):

$R = 2n/(N + N')$, using the notation presented above. In fact, we use a somewhat more complicated definition that accounts for the quality of spatial superposition of matched atoms and their distance from X′."

We do not want to discuss here these definitions for these factors and states. Our purpose is not the demonstration of effectiveness of these definitions or effectiveness of QSAR. The interested reader can learn it from papers [3, 4] and references inside of these papers. As we said above,

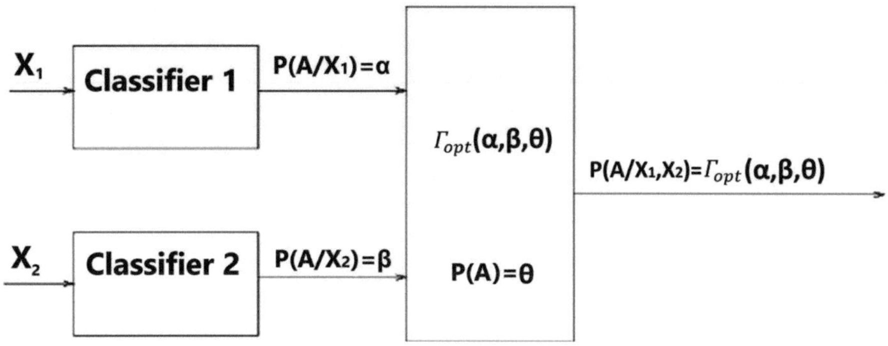

Figure 2. Function $\Gamma_{opt}(\alpha, \beta, \theta):[0,1]^3 \rightarrow [0,1]$.

we use QSAR only for clearness; the proof is correct for any field of use of Naive Bayes classifier.

Let us consider the case when no relationships exist between reliabilities. In this case, the Naive Bayes model is a correct arrangement of the issue. We demonstrate in this chapter that for the case that we don't know relationships between reliabilities even approximately—the Naive Bayes model is not correct, but ideal arrangement in a few senses. More point by point, we demonstrate that the Naive Bayes model gives minimal mean error over all conceivable models of relationship. We assume in this confirmation that all relationship models have the same likelihood. We think that this result can clarify the depicted over secretive optimality of Naive Bayes model.

The Chapter is built as described in the following statements. We grant correct numerical description of the issue for two states and two reliabilities in Section 2. We characterize our notations in Section 3. We define general form of conditional likelihood for all conceivable relationships of our reliabilities in Section 4. We characterize the limitations of the functions depicting the relationships in Section 5. We find the formula for an interval between two models of probability (correlation) in Section 6. We discover constraints for our fundamental functions in Section 7. We illuminate our primary issue; we demonstrate Naive Bayes model's optimality for uniform distribution of all conceivable relationships in Section 8. We discover mean error between the Naive Bayes model and a genuine model for uniform distribution of all conceivable relationships in Section 9. We consider the case of more than two states and reliabilities in Section 10. We make conclusions in Section 11.

2. Definition of the task

Suppose that A is a state for "ligand-active site of protein" complex. It is 0 (ligand is not bound to active site of protein) or 1 (ligand is not bound to active site of protein). Accept that the *apriori* likelihood $P(A) = P(A = 1)$ is known, and indicate it by θ. Let X_1, X_2 be two reliability values (defined above), with values in a set $[0; 1]$. However, for generality, we will define X_1, X_2 in a set $[-\infty;+\infty]$, but probability density to find X_1, X_2 in $[-\infty; 0]$ or $[1;+\infty]$ is equal to zero. We have the taking after data: $X_1 = x_1$ and $X_2 = x_2$ (gotten through estimation). Moreover, we have two functions, "classifiers," which for given x_1 and x_2 give us

$$P(A = 1/X_1 = x_1) = P(A/x_1) \equiv \alpha,$$

$$P(A = 1/X_2 = x_2) = P(A/x_2) \equiv \beta.$$

We want to find the likelihood

$$P(A = 1/X_1 = x_1, X_2 = x_2) = P(A/x_1, x_2)$$

in terms of α, β and θ. More particularly we wish to discover a function $\Gamma_{opt}(\alpha, \beta, \theta)$ which on the average is the most excellent estimation for $P(A/x_1, x_2)$ in a sense to be characterized expressly within the following consideration (see **Figure 2**).

3. Notation and preliminaries

$\rho_{X_1, X_2}(x_1, x_2)$—joint PDF (probability density function) for X_1 and X_2.

$\rho_{X_1, X_2/A}(x_1, x_2) = h(x_1, x_2)$—joint PDF for X_1 and X_2, known $A = 1$. In terms of $h(x_1, x_2)$ and θ, we can find $P(A/x_1, x_2)$ as follows:

$$P(A/x_1, x_2) = \frac{\theta h(x_1, x_2)}{\theta h(x_1, x_2) + (1 - \theta)\overline{h}(x_1, x_2)}, \tag{1}$$

here

$\overline{h}(x_1, x_2) \equiv \rho_{X_1, X_2/\overline{A}}(x_1, x_2)$—joint PDF for X_1 and X_2, known $A = 0$.

We can find

$$\rho_{X_1}(x_1) = \int_{-\infty}^{+\infty} \rho_{X_1, X_2}(x_1, x_2)dx_2,$$

$$\rho_{X_2}(x_2) = \int_{-\infty}^{+\infty} \rho_{X_1, X_2}(x_1, x_2)dx_1,$$

$$h_1(x_1) \equiv \rho_{X_1/A}(x_1) = \int_{-\infty}^{+\infty} h(x_1, x_2)dx_2,$$

$$h_2(x_2) \equiv \rho_{X_2/A}(x_2) = \int_{-\infty}^{+\infty} h(x_1, x_2)dx_1,$$

$$\overline{h}_1(x_1) \equiv \rho_{X_1/\overline{A}}(x_1) = \int_{-\infty}^{+\infty} \overline{h}(x_1, x_2) dx_2,$$

$$\overline{h}_2(x_2) \equiv \rho_{X_2/\overline{A}}(x_2) = \int_{-\infty}^{+\infty} \overline{h}(x_1, x_2) dx_1.$$

4. Generic form of $P(A/x_1, x_2)$

Let us define the function $g(x_1, x_2)$ and $\overline{g}(x_1, x_2)$

$$g(x_1, x_2) \equiv \frac{h(x_1, x_2)}{h_1(x_1) h_2(x_2)},$$

$$\overline{g}(x_1, x_2) \equiv \frac{\overline{h}(x_1, x_2)}{\overline{h}_1(x_1) \overline{h}_2(x_2)}.$$

Let us say that if X_1 and X_2 are conditionally independent, i.e.,

$$\begin{aligned} h(x_1, x_2) &= \rho_{X_1 X_2/A}(x_1, x_2) = \rho_{X_1/A}(x_1) \rho_{X_2/A}(x_2) \\ &= h_1(x_1) h_2(x_2), \end{aligned}$$

then

$$g(x_1, x_2) = \overline{g}(x_1, x_2) = 1.$$

Let us define the following *monotonously nondecreasing* probability distribution functions:

$$H_1(x_1) \equiv \int_{-\infty}^{x_1} h_1(z) dz,$$

$$H_2(x_2) \equiv \int_{-\infty}^{x_2} h_2(z) dz,$$

$$\overline{H}_1(x_1) \equiv \int_{-\infty}^{x_1} \overline{h}_1(z) dz,$$

$$\overline{H}_2(x_2) \equiv \int_{-\infty}^{x_2} \overline{h}_2(z) dz.$$

Take attention that since $H_1(x_1), H_2(x_2), \overline{H}_1(x_1)$ and $\overline{H}_2(x_2)$ are monotonous (At this point we can assume that $h_1(x_1), h_2(x_2), \overline{h}_1(x_1), \overline{h}_2(x_2) > 0$ so that $H_1(x_1), H_2(x_2), \overline{H}_1(x_1)$ and $\overline{H}_2(x_2)$ are monotonously increasing. However, such limitation will be unnecessary as we will see within the following conclusion.), the inverse functions $H_1^{-1}(x_1), H_2^{-1}(x_2), \overline{H}_1^{-1}(x_1)$ and $\overline{H}_2^{-1}(x_2)$ must exist. As a result, we can characterize

$$J(a,b) \equiv g\left(H_1^{-1}(a), H_2^{-1}(b)\right),$$

$$\overline{J}(a,b) \equiv \overline{g}\left(\overline{H}_1^{-1}(a), \overline{H}_2^{-1}(b)\right).$$

To be brief, let us use the following concise designation:

$$J \equiv J(H_1(x_1), H_2(x_2)) = g\left(H_1^{-1}(H_1(x_1))\right), H_2^{-1}(H_2(x_2)) = g(x_1, x_2),$$

$$\overline{J} \equiv \overline{J}\left(\overline{H}_1(x_1), \overline{H}_2(x_2)\right) = \overline{g}\left(\overline{H}_1^{-1}\left(\overline{H}_1(x_1)\right), \overline{H}_2^{-1}\left(\overline{H}_2(x_2)\right)\right) = \overline{g}(x_1, x_2).$$

By the definition

$$h(x_1, x_2) = J h_1(x_1) h_2(x_2), \tag{2}$$

$$\overline{h}(x_1, x_2) = \overline{J}\, \overline{h}_1(x_1) \overline{h}_2(x_2). \tag{3}$$

We currently obtain

$$h_1(x_1) \equiv \rho_{X_1/A}(x_1) = \frac{\rho_{X_1}(x_1) P(A/x_1)}{P(A)} = \frac{\alpha \rho_{X_1}(x_1)}{\theta}, \tag{4}$$

$$h_2(x_2) \equiv \rho_{X_2/A}(x_2) = \frac{\rho_{X_2}(x_2) P(A/x_2)}{P(A)} = \frac{\beta \rho_{X_2}(x_2)}{\theta}, \tag{5}$$

$$\overline{h}_1(x_1) \equiv \rho_{X_1/\overline{A}}(x_1) = \frac{\rho_{X_1}(x_1) P\left(\overline{A}/x_1\right)}{P\left(\overline{A}\right)} = \frac{(1-\alpha) \rho_{X_1}(x_1)}{1-\theta}, \tag{6}$$

$$\overline{h}_2(x_2) \equiv \rho_{X_2/\overline{A}}(x_2) = \frac{\rho_{X_2}(x_2) P\left(\overline{A}/x_2\right)}{P\left(\overline{A}\right)} = \frac{(1-\alpha) \rho_{X_2}(x_2)}{1-\theta}. \tag{7}$$

As a result from Eqs. (2) and (3)

$$h(x_1, x_2) = J \frac{\alpha \beta \rho_{X_1}(x_1) \rho_{X_2}(x_2)}{\theta^2},$$

$$\overline{h}(x_1, x_2) = \overline{J} \frac{(1-\alpha)(1-\beta) \rho_{X_1}(x_1) \rho_{X_2}(x_2)}{(1-\theta)^2}.$$

Now from Eq. (1)

$$P(A/x_1,x_2) = \frac{\frac{J}{\theta}\alpha\beta\rho_{X_1}(x_1)\rho_{X_2}(x_2)}{\frac{J}{\theta}\alpha\beta\rho_{X_1}(x_1)\rho_{X_2}(x_2) + \frac{\bar{J}}{(1-\theta)}(1-\alpha)(1-\beta)\rho_{X_1}(x_1)\rho_{X_2}(x_2)}$$
$$= \frac{\alpha\beta}{\alpha\beta + \frac{\bar{J}}{J}\frac{\theta}{1-\theta}(1-\alpha)(1-\beta)}. \tag{8}$$

Note, that for values of $J = \bar{J} = 1$ (conditional independence of x_1 and x_2) equation (8) becomes the exact solution for the optimal model:

$$\Gamma(\alpha,\beta,\theta) = P(A/x_1,x_2).$$

5. Limitations for the functions $J(a,b)$ and $\bar{J}(a,b)$

We can write

$$h_1(x_1) = \int_{-\infty}^{+\infty} J(H_1(x_1),H_2(x_2))h_1(x_1)h_2(x_2)dx_2. \tag{9}$$

As a result

$$1 = \int_{-\infty}^{+\infty} J(H_1(x_1),H_2(x_2))h_2(x_2)dx_2 = \int_0^1 J(H_1(x_1),H_2(x_2))dH_2(x_2). \tag{10}$$

Thus, we obtain the following condition:

$$\int_0^1 J(a,b)db = 1, \tag{11}$$

and similarly

$$\int_0^1 J(a,b)da = 1. \tag{12}$$

Similarly, we can get

$$\int_0^1 \bar{J}(a,b)da = 1$$
$$\int_0^1 \bar{J}(a,b)db = 1. \tag{13}$$

Obviously

$$J(a,b), \bar{J}(a,b) \geq 0, \tag{14}$$

$$\int_0^1\int_0^1 J(a,b)dadb = \int_0^1\int_0^1 \bar{J}(a,b)dadb = 1. \tag{15}$$

All the solutions of Eqs. (11)–(15) together with (8) can define the set of all possible realizations of $P(A/x_1, x_2)$.

Let us give some example of a solution of (11), (12) and (14), (15):

Let $\rho(x)$ be a function such that $\rho(x) \geq 0$ and $\int_0^1 \rho(x)dx = 1$

$$J(a,b) = \begin{cases} \rho(a-b) & , a \geq b \\ \rho(a-b+1) & , a < b' \end{cases}$$

satisfy Eqs. (11), (12), (14), and (15).

6. Definition of distance

We define the distance between the proposed approximation of $P(A/x_1, x_2)$-$\Gamma(\alpha, \beta, \theta)$ and the actual function $P(A/x_1, x_2)$ as follows:

$$\|\Gamma(\alpha,\beta,\theta) - P(A/x_1,x_2)\| = \int\int_{-\infty}^{+\infty} \rho_{X_1X_2}(x_1,x_2)\left[\Gamma(\alpha,\beta,\theta) - P(A/x_1,x_2)\right]^2 dx_1dx_2.$$

Now we have from Eqs. (2) and (3) and Eqs. (4)–(7)

$$\begin{aligned} \rho_{X_1X_2}(x_1,x_2) &= \theta h(x_1,x_2) + (1-\theta)\bar{h}(x_1,x_2) \\ &= \theta J h_1(x_1)h_2(x_2) + (1-\theta)\bar{J}\bar{h}_1(x_1)\bar{h}_2(x_2) \\ &= \left[J\frac{\alpha\beta}{\theta} + \bar{J}\frac{(1-\alpha)(1-\beta)}{(1-\theta)}\right]\rho_{X_1}(x_1)\rho_{X_2}(x_2), \end{aligned}$$

$$\|\Gamma(\alpha,\beta,\theta)-P(A/x_1,x_2)\| = \int_{-\infty}^{+\infty}\!\!\int \rho_{X_1}(x_1)\rho_{X_2}(x_2)\left[J\frac{\alpha\beta}{\theta}+\bar{J}\frac{(1-\alpha)(1-\beta)}{(1-\theta)}\right]\times(\Gamma(\alpha,\beta,\theta)-P(A/x_1,x_2))^2dx_1dx_2$$
$$=\int_0^1\!\!\int_0^1\left[\frac{J\alpha\beta}{\theta}+\frac{\bar{J}(1-\alpha)(1-\beta)}{(1-\theta)}\right]\times(\Gamma(\alpha,\beta,\theta)-P(A/x_1,x_2))^2dF_1(x_1)dF_2(x_2). \quad (16)$$

Here

$$F_1(x_1)=\int_{-\infty}^{x_1}\rho_{X_1}(z)dz,$$

$$F_2(x_2)=\int_{-\infty}^{x_2}\rho_{X_2}(z)dz.$$

7. Constraints for basic functions

We will consider further all functions with arguments $1\geq F_1, F_2\geq 0$, but not x_1, x_2. We have six functions of F_1, F_2, which define Eq. (16): $J, \bar{J}, H_1, H_2, \alpha, \beta$. Let us to write the functions by help these functions (F_1, F_2) and find restrictions for these functions:

$$\alpha=P(A/x_1)=\theta h_1(x_1)/\rho_{X_1}(x_1)=\theta(dH_1/dx_1)/(dF_1/dx_1)=\theta\frac{dH_1}{dF_1}.$$

By the same way

$$\beta=\theta\frac{dH_2}{dF_2}.$$

We know that functions H_1, F_1, H_2, F_2 are cumulative distribution functions of x_1,x_2, correspondingly. These functions are *monotonously nondecreasing* and change from 0 to 1 from the definition of cumulative distribution functions. Therefore, we can conclude the following restraints for functions H_1, H_2 as functions of F_1, F_2 exist:

$$H_1(1)=H_2(1)=1,$$
$$H_1(0)=H_2(0)=0,$$

$$0\leq\alpha=\theta\frac{dH_1}{dF_1},\beta=\theta\frac{dH_2}{dF_2}\leq 1,$$

$$0\leq\theta\leq 1,$$

$$\begin{aligned}\overline{H}_1(x_1) &= \int_{-\infty}^{x_1} \overline{h}_1(x_1)dx_1 \\ &= \int_{-\infty}^{x_1} \frac{(1-\alpha)\rho_{X_1}(x_1)}{1-\theta}dx_1 \\ &= \frac{1}{1-\theta}\int_{-\infty}^{x_1} \rho_{X_1}(x_1)dx_1 - \frac{\theta}{1-\theta}\int_{-\infty}^{x_1} \frac{\alpha\rho_{X_1}(x_1)}{\theta}dx_1 \\ &= \frac{F_1}{1-\theta} - \frac{\theta}{1-\theta}H_1(x_1).\end{aligned}$$

By the same way

$$\overline{H}_2(x_2) = \frac{F_2}{1-\theta} - \frac{\theta}{1-\theta}H_2(x_2),$$

$$J(H_1(F_1), H_2(F_2)) : J(H_1(F_1), H_2(F_2)) \geq 0$$

$$\int_0^1 J(a,b)db = 1$$

$$\int_0^1 J(a,b)da = 1,$$

$$\overline{J}(\overline{H}_1(F_1), \overline{H}_2(F_2)) : \overline{J}(\overline{H}_1(F_1), \overline{H}_2(F_2)) \geq 0$$

$$\int_0^1 \overline{J}(a,b)db = 1$$

$$\int_0^1 \overline{J}(a,b)da = 1,$$

$$P(A/x_1, x_2) = \frac{\frac{J\alpha\beta}{\theta}}{\frac{J\alpha\beta}{\theta} + \frac{\overline{J}(1-\alpha)(1-\beta)}{1-\theta}}. \tag{17}$$

8. Optimization

We shall find the best approximation of $\Gamma(\alpha, \beta, \theta)$ as follows:

$$min_{\Gamma(\alpha,\beta,\theta)} E[\|\Gamma(\alpha,\beta,\theta) - P(A/x_1,x_2)\|] \rightarrow \Gamma(\alpha,\beta,\theta),$$

where the expected value (or expectation or mathematical expectation or mean or the first moment) $E[\ldots]$ is taken with respect to the joint PDF of possible realizations of $J, \overline{J}, \alpha, \beta, H_1, H_2$ for given F_1 and F_2.

For the sake of brevity, we denote

$$C \equiv \frac{J\alpha\beta}{\theta} + \frac{\overline{J}(1-\alpha)(1-\beta)}{1-\theta}$$

$$D \equiv \frac{J\alpha\beta}{\theta}.$$

Then from Eqs. (17) and (16)

$$\begin{aligned}\|\Gamma(\alpha,\beta,\theta)-P(A/x_1,x_2)\| &= \int_0^1\int_0^1 C(\Gamma(\alpha,\beta,\theta)-D/C)^2 dF_1 dF_2\\ &= \int_0^1\int_0^1 dF_1 dF_2\left[D^2C+\Gamma^2(\alpha,\beta,\theta)C-2\Gamma(\alpha,\beta,\theta)D\right].\end{aligned}\tag{18}$$

Thus

$$\begin{aligned}& min_{\Gamma(\alpha,\beta,\theta)}E\left[\|\Gamma(\alpha,\beta,\theta)-P(A/x_1,x_2)\|\right]\\ &= min_{\Gamma(\alpha,\beta,\theta)}E\left[\int_0^1\int_0^1 dF_1dF_2\left[D^2C+\Gamma^2(\alpha,\beta,\theta)C-2\Gamma(\alpha,\beta,\theta)D\right]\right]\\ &= min_{\Gamma(\alpha,\beta,\theta)}E\left[\int_0^1\int_0^1 dF_1dF_2\left[D^2C\right]\right]\\ &\quad + min_{\Gamma(\alpha,\beta,\theta)}E\left[\int_0^1\int_0^1 dF_1dF_2\left[\Gamma^2(\alpha,\beta,\theta)C-2\Gamma(\alpha,\beta,\theta)D\right]\right]\\ &= Const + min_{\Gamma(\alpha,\beta,\theta)}E\left[\int_0^1\int_0^1 dF_1dF_2\left[\Gamma^2(\alpha,\beta,\theta)C-2\Gamma(\alpha,\beta,\theta)D\right]\right].\end{aligned}\tag{19}$$

It remains to calculate the expected value in Eq. (19).

We have by obvious assumptions

$$\begin{aligned}\rho_{J,\bar{J},\alpha,\beta,H_1H_2/F_1,F_2}(J,\bar{J},\alpha,\beta,H_1,H_2/F_1,F_2) &= \rho_{J/H_1,H_2}(J/H_1,H_2)\rho_{\bar{J}/\overline{H}_1,\overline{H}_2}(\bar{J}/\overline{H}_1,\overline{H}_2)\rho_{\alpha/F_1}(\alpha/F_1)\\ &\quad\times\rho_{H_1/\alpha F_1}(H_1/\alpha,F_1)\rho_{\beta/F_2}(\beta/F_2)\rho_{H_2/\beta,F_2}(H_2/\beta,F_2).\end{aligned}\tag{20}$$

Lemma 1

$$E[J(a,b)] = \int_0^{+\infty}\rho_{J(a,b)/a,b}(J(a,b)/a,b)J(a,b)dJ = 1,$$

$$E\left[\bar{J}(a,b)\right] = \int_0^{+\infty}\rho_{\bar{J}(a,b)/a,b}(\bar{J}(a,b)/a,b)\bar{J}(a,b)d\bar{J} = 1.$$

Proof: We can take into the consideration the function $\rho_{J(a,b)/a,b}$. The domain of the function $J(a,b)$ is square $0 \le a,b \le 1$. By dividing this square into small squares (i,j), we can get sampling of the function J. Then, on every square i,j, we can define the value of the function J_{ij}. We can write the following restraints for function $J(***)$:

$$J_{ij} \geq 0,$$

$$\frac{1}{N}\sum_{i=1}^{N} J_{ij} = 1,$$

$$\frac{1}{N}\sum_{j=1}^{N} J_{ij} = 1.$$

Here $i = 1, \ldots, N$, $j = 1, \ldots, N$.

All matrixes $\left(J_{ij}\right)$ that satisfy the above conditions have the same probability. So we can define probability density function

$$\rho\left(J_{11}, \ldots, J_{ij}, \ldots, J_{NN}\right).$$

This density function should be symmetric according to transpositions of columns and rows of the matrix $\left(J_{ij}\right)$, because the density function has the same probability for all matrixes $\left(J_{ij}\right)$ that satisfy the above conditions. Indeed, these conditions are also symmetric according to transpositions of columns and rows of matrix $\left(J_{ij}\right)$. From symmetry conditions that define this function (ρ) according to transpositions of columns and rows of matrix $\left(J_{ij}\right)$, it is possible to conclude that this function (ρ) also does not transform according to these transpositions.

We can consider function $\rho_{u/ij}(u/ij)$ that is a discretization of the function $\rho_{J(a,b)/a,b}(J(a,b)/a,b)$:

$$\rho_{u/ij}(u/ij) = \int \ldots \int_0^{+\infty} \rho\left(J_{11}, \ldots J_{nk}, \ldots J_{ij} = u, \ldots, J_{NN}\right) \prod\nolimits_{(lm) \neq (ij)} dJ_{lm}.$$

We can transpose columns and rows $\left(J_{ij}\right)$ in such a way that element J_{ij} will be replaced by the other element J_{nk}, and after it the function $\rho(J_{11}, \ldots)$ will not be transformed. So from the above equation, we can get

$$\rho_{u/ij}(u/ij) = \int \ldots \int_0^{+\infty} \rho\left(J_{11}, \ldots J_{nk}, \ldots J_{ij} = u, \ldots, J_{NN}\right) \prod\nolimits_{(lm) \neq (nk)} dJ_{lm} = \rho_{u/nk}(u/nk).$$

From this equation we can conclude that $\rho_{u/ij}(u/ij)$ does not depend on ij so $\rho_{J/a,b}(J/a,b)$ does not depend on ab and

$$\rho_{J/a,b}(J/a,b) = \rho_J(J),$$

and

$$E[J(a,b)] = \int_0^{+\infty} \rho_J(J) J dJ = Const, \tag{21}$$

From

$$\int_0^1 \int_0^1 J(a,b)dadb = 1,$$

we can conclude that

$$\int_0^1 \int_0^1 E[J(a,b)]dadb = 1.$$

So we can obtain that $Const = 1$ in Eq. (21).

Lemma 2: Probability distribution functions α and β do not depend on F_1 and F_2:

$$\rho_{\alpha/F_1}(\alpha/F_1) = \rho_\alpha(\alpha),$$

$$\rho_{\beta/F_2}(\beta/F_2) = \rho_\beta(\beta).$$

Proof: Let us make sampling of the function $\alpha(F_1)$ by dividing the domain of this function F_1, $[0, 1]$ on intervals of $1/N, N \gg 1$. Then restriction conditions for α_k, $k = 1, \ldots, N$

$$0 \le \alpha_k \le 1,$$

$$\frac{1}{N}\sum_{k=1}^{N} \alpha_k = \int_0^1 \theta dH_1(F_1)dF_1 dF_1 = \theta.$$

All columns (α_k) that are satisfied by these conditions have equal probability. We can consider respective function $\rho(\alpha_1, \ldots, \alpha_k, \ldots, \alpha_l, \ldots, \alpha_N)$. From symmetry conditions that define this function according to transpositions $\alpha_k \to \alpha_l$, function $\rho(\alpha_1, \ldots, \alpha_k, \ldots, \alpha_l, \ldots, \alpha_N)$ also does not transform according to such transpositions. As a result, it is possible to write

$$\begin{aligned}\rho_k(u) &= \int_0^1 \rho(\alpha_1, \ldots, \alpha_k = u, \ldots, \alpha_l, \ldots, \alpha_N) \prod\nolimits_{n \neq k} d\alpha_n \\ &= \int_0^1 \rho(\alpha_1, \ldots, \alpha_k, \ldots, \alpha_l = u, \ldots, \alpha_N) \prod\nolimits_{n \neq l} d\alpha_n \\ &= \rho_l(u).\end{aligned}$$

From this equation, we can conclude that function $\rho_{\alpha/F_1}(\alpha/F_1)$ does not depend on F_1:

$$\rho_{\alpha/F_1}(\alpha/F_1) = \rho_\alpha(\alpha).$$

From (20) we obtain

$$E\left[\Gamma^2(\alpha,\beta,\theta)C - 2\Gamma(\alpha,\beta,\theta)D\right] = \int_0^1\int_0^1 \rho_\alpha(\alpha)\rho_\beta(\beta)d\alpha d\beta$$

$$\times \int_0^1\int_0^1 \rho_{H_1/\alpha,F_1}(H_1/\alpha,F_1)\rho_{H_2/\beta,F_2}(H_2/\beta,F_2)dH_1dH_2\int_0^\infty\int_0^\infty \rho_J(J)\rho_{\bar{J}}(\bar{J})$$

$$\times\left[\Gamma^2(\alpha,\beta,\theta)\left[\frac{J\alpha\beta}{\theta}+\frac{\bar{J}(1-\alpha)(1-\beta)}{1-\theta}\right] - 2\Gamma(\alpha,\beta,\theta)\frac{J\alpha\beta}{\theta}\right]dJd\bar{J}$$

$$= \int_0^1\int_0^1 \rho_\alpha(\alpha)\rho_\beta(\beta)d\alpha d\beta$$

$$\times\left[\Gamma^2(\alpha,\beta,\theta)\left[\frac{E[J]\alpha\beta}{\theta}+\frac{E\left[\bar{J}\right](1-\alpha)(1-\beta)}{1-\theta}\right] - 2\Gamma(\alpha,\beta,\theta)\frac{E[J]\alpha\beta}{\theta}\right].$$

Let us define

$$\overline{C} = \frac{\alpha\beta}{\theta}+\frac{(1-\alpha)(1-\beta)}{1-\theta},$$

$$\overline{D} = \frac{\alpha\beta}{\theta}.$$

By Lemma 1, $E[J] = E\left[\bar{J}\right] = 1$. Hence

$$E\left[\Gamma^2(\alpha,\beta,\theta)C - 2\Gamma(\alpha,\beta,\theta)D\right] = \int_0^1\int_0^1 \left[\Gamma^2(\alpha,\beta,\theta)\overline{C} - 2\Gamma(\alpha,\beta,\theta)\overline{D}\right]\rho_\alpha(\alpha)\rho_\beta(\beta)d\alpha d\beta.$$

It remains to find

$$min_{\Gamma(\alpha,\beta,\theta)}\int_0^1\int_0^1 dF_1dF_2\int_0^1\int_0^1 d\alpha d\beta\rho_\alpha(\alpha)\rho_\beta(\beta)\left[\Gamma^2(\alpha,\beta,\theta))\overline{C} - 2\Gamma(\alpha,\beta,\theta)\overline{D}\right]. \tag{22}$$

Since

$$\rho_\alpha(\alpha)\rho_\beta(\beta) \geq 0,$$

if the expression in square brackets is minimized at each point, then the whole integral in Eq. (22) is minimized. Thus, we may proceed as follows:

$$\frac{\partial}{\partial \Gamma}\left[\Gamma^2(\alpha,\beta,\theta)\overline{C} - 2\Gamma(\alpha,\beta,\theta)\overline{D}\right] = 2\Gamma(\alpha,\beta,\theta)\overline{C} - 2\overline{D} = 0.$$

Hence the optimum $\Gamma(\alpha,\beta,\theta)$ is given by

$$\Gamma_{opt}(\alpha,\beta,\theta) = \frac{\overline{D}}{\overline{C}} = \frac{\frac{\alpha\beta}{\theta}}{\frac{\alpha\beta}{\theta} + \frac{(1-\alpha)(1-\beta)}{1-\theta}}.$$

9. Mean distance between the proposed approximation of $P(A/x_1,x_2) - \Gamma(\alpha,\beta,\theta)$ and the actual function $P(A/x_1,x_2)$

The mean distance from (18) is

$$\begin{aligned} DIS &= E\left[\left\|\Gamma(\alpha,\beta,\theta) - P(A/x_1,x_2)\right\|\right] \\ &= \int_0^1 \int_0^1 \rho_\alpha(\alpha)\rho_\beta(\beta)d\alpha d\beta\left[\Gamma^2(\alpha,\beta,\theta)\overline{C} - 2\Gamma(\alpha,\beta,\theta)\overline{D}\right] + Const, \end{aligned}$$

where *Const* in this equation is defined by

$$Const = E\left[\int_{-\infty}^{+\infty}\int_{-\infty}^{+\infty} \rho_{X_1X_2}(x_1,x_2)[P(A/x_1,x_2)]^2 dx_1dx_2\right].$$

From this equation we can find boundaries of the *Const*. From $0 \le P(A/x_1,x_2) \le 1$ we can conclude

$$Const \le E\left[\int_{-\infty}^{+\infty}\int_{-\infty}^{+\infty} \rho_{X_1,X_2,}(x_1,x_2)P(A/x_1,x_2)dx_1dx_2\right] = E[\theta] = \theta.$$

The second condition is

$$\begin{aligned} 0 &\le E\left[\int_{-\infty}^{+\infty}\int_{-\infty}^{+\infty} \rho_{X_1,X_2}(x_1,x_2)[P(A/x_1,x_2) - \theta]^2 dx_1dx_2\right] \\ &= E\left[\int_{-\infty}^{+\infty}\int_{-\infty}^{+\infty} \rho_{X_1,X_2}(x_1,x_2)\left[P(A/x_1,x_2)^2 + \theta^2 - 2P(A/x_1,x_2)\theta\right]dx_1dx_2\right] \\ &= E\left[\int_{-\infty}^{+\infty}\int_{-\infty}^{+\infty} \rho_{X_1,X_2}(x_1,x_2)[P(A/x_1,x_2)]^2 dx_1dx_2\right] - \theta^2. \end{aligned}$$

So from these two equations, we can conclude

$$\theta^2 \le Const \le \theta.$$

In the next step, we would like find function $\rho_\alpha(\alpha)$ $(\rho_\beta(\beta))$ in the equation for *DIS*.

Restrictions for function $\alpha(F_1), 0 \le F_1 \le 1$ are the following:

$$\int_0^1 \alpha(F_1)dF_1 = \theta,$$

$$0 \le \alpha(F_1) \le 1.$$

In discrete form (for $N \to \infty$), we can rewrite $\alpha_{set} = \{\alpha_1, \alpha_2, \ldots, \alpha_N\}$

$$\frac{1}{N}\sum_{i=1}^{N} \alpha_i = \theta,$$

$$0 \le \alpha_i \le 1, i = 1, 2, \ldots, N.$$

Let us define a function $U(\alpha_{set})$ in the following way:

$$U(\alpha_{set}) = \begin{cases} \sum_{i=1}^{N} \alpha_i & \text{for } 0 \le \alpha_i \le 1 \quad , i = 1, 2, \ldots, N \\ +\infty & \text{otherwise} \end{cases},$$

$$U(\alpha_{set}) = \sum_{i=1}^{N} U_i(\alpha_i),$$

$$U_i(\alpha_i) = \begin{cases} \alpha_i & \text{for } 0 \le \alpha_i \le 1 \\ +\infty & \text{otherwise} \end{cases}.$$

Then the function that satisfies equal probability distribution with considering restrictions (i) and (ii) is the following:

$$\rho_{\alpha_{set}}(\alpha_{set}) = \frac{1}{C}\delta(U(\alpha_{set}) - N\theta). \tag{23}$$

here δ is the Dirac delta function.

We can define the constant C by

$$\int_{-\infty}^{+\infty} \cdots \int_{-\infty}^{+\infty} \rho_{\alpha_{set}}(\alpha_{set}) d\alpha_1 \ldots d\alpha_N = 1.$$

It can be proved for $N \to \infty$ that distribution (23) is equal to the following distribution (from "statistical mechanics" [9]; transform from microcanonical to canonical distribution):

$$\rho_{\alpha_{set}}(\alpha_{set}) = \frac{1}{Z} e^{-KU(\alpha_{set})}.$$

Here we can find Z and K from the following equations:

$$\int_{-\infty}^{+\infty} \dots \int_{-\infty}^{+\infty} \rho_{\alpha_{set}}(\alpha_{set}) d\alpha_1 \dots d\alpha_N = 1, \tag{24}$$

$$\int_{-\infty}^{+\infty} \dots \int_{-\infty}^{+\infty} U(\alpha_{set}) \rho_{\alpha_{set}}(\alpha_{set}) d\alpha_1 \dots d\alpha_N = N\theta. \tag{25}$$

Quest function $\rho_\alpha(\alpha)$ can be found by

$$\rho_\alpha(\alpha) = \int_{-\infty}^{+\infty} \dots \int_{-\infty}^{+\infty} \rho_{\alpha_{set}}(\alpha_1, \dots, \alpha_j = \alpha, \dots, \alpha_N) \prod_{i=1,\, i \neq j}^{N} d\alpha_i = \frac{1}{D} e^{-KU_j(\alpha_j = \alpha)}, \tag{26}$$

where

$$D^N = Z. \tag{27}$$

From Eqs. (24) and (25), we can find

$$\frac{1}{Z} = \left(\frac{K}{1 - e^{-K}} \right)^N, \tag{28}$$

$$\theta = \Lambda(K), \tag{29}$$

where $\Lambda(K)$ is the decreasing function

$$\Lambda(K) = \begin{cases} 1 & \text{for } K = -\infty \\ 0 & \text{for } K = +\infty \\ 1/2 & \text{for } K = 0 \\ \dfrac{1}{K} - \dfrac{1}{e^K - 1} & \text{otherwise} \end{cases}.$$

If K is the root of Eq. (29), we can write from Eqs. (26)–(29) for function $\rho_\alpha(\alpha)$

$$\rho_\alpha(\alpha) = \begin{cases} \begin{cases} \text{For} & K = 0 \\ 1 & \text{for } 0 \le \alpha \le 1 \\ 0 & \text{otherwise } \alpha \end{cases} \\ \begin{cases} \text{For} & K = +\infty \\ 2\delta(\alpha) & 0 \le \alpha \le 1 \\ 0 & \text{otherwise } \alpha \end{cases} \\ \begin{cases} \text{For} & K = -\infty \\ 2\delta(\alpha - 1) & 0 \le \alpha \le 1 \\ 0 & \text{otherwise } \alpha \end{cases} \\ \begin{cases} \text{For} & \text{otherwise } K \\ \frac{1}{D} e^{-K\alpha} & 0 \le \alpha \le 1 \\ 0 & \text{otherwise } \alpha \end{cases} \end{cases},$$

where $2\int_0^1 \delta(\alpha - 1) d\alpha = 2\int_0^1 \delta(\alpha) d\alpha = 1$ and $\frac{1}{D} = \frac{K}{1-e^{-K}}$.

10. The case of more than two states A and reliabilities X

Let A be a state, with values in set 0, 1, …, L. This number can characterize strength of a bond. Assume that the *apriori* probability $P(A = i)$ is known, and denote it by θ_i; here $i = 1, \ldots, L$. Let $X_1, \ldots, X_K$ be random variables, with values in some set, say $]-\infty; +\infty[$. We have the following information: $X_1 = x_1, \ldots, X_K = x_K$ (obtained through measurement). Furthermore, we have systems, "classifiers," which for given $x_1, \ldots, x_K$ produce

$$P(A = i/X_j = x_j) \equiv \alpha_{ij}.$$

We want to find the probability $P(A = i/X_1 = x_1, \ldots, X_K = x_K)$ in terms of α_{ij} and θ_i. In more detail, we want to find a function $\Gamma_{opt,M}(\alpha_{ij}, \theta_i)$, which is the best approximation for $P(A = M/x_1, \ldots, x_K)$ on the average. By the same way, in case of two variables, it is possible to find that the $\Gamma_{opt,M}(\alpha_{ij}, \theta_i)$ can be defined by the following equation:

$$\Gamma_{opt,M}(\alpha_{ij}, \theta_i) = \frac{\left(\prod_{j=1}^{K} \alpha_{Mj}\right)/\theta_M^{K-1}}{\sum_{i=1}^{L} \left(\prod_{j=1}^{K} \alpha_{ij}\right)/\theta_i^{K-1}}.$$

We have evidential restraints for α_{ij}, θ_i

$$0 \le \alpha_{ij} \le 1$$
$$\sum_{i=1}^{L} \alpha_{ij} = 1,$$
$$0 \le \theta_i \le 1$$
$$\sum_{i=1}^{L} \theta_i = 1.$$

11. Conclusions

Using as an illustration the QSAR, we demonstrated effectively that the Naive Bayes model gives minimal mean error over uniform dispersion of all conceivable relationships between characteristic reliabilities. This result can clarify the portrayed over secretive optimality of Naive Bayes model. We too found the mean error that the Naive Bayes model gives for uniform distribution of all conceivable relationships of reliabilities.

Medicinal chemistry (quantitative structure-activity relationships, QSAR) prediction increasingly relies on Bayesian network-based methods. Its importance derives partly from the difficulty and inaccuracies of present quantum chemical models (e.g., in SYBYL and other software) and from the impracticality of sufficient characterization of structure of drug molecules and receptor active sites, including vicinal waters in and around hydrophobic pockets in active sites. This is particularly so for biologicals (protein and nucleic acid APIs (nucleic acid active pharmaceutical ingredients)) and target applications that exhibit extensive inter-receptor trafficking, genomic polymorphisms, and other system biology phenomena. The effectiveness and accuracy of Bayesian methods for drug development likewise depend on certain prerequisites, such as an adequate distance metric by which to measure similarity/difference between combinatorial library molecules and known successful ligand molecules targeting a particular receptor and addressing a particular clinical indication. In this connection, the distance metric proposed in Section 6 of the chapter manuscript and the associated Lemmas and Proofs are of substantial value in the future of high-throughput screening (HTS) and medicinal chemistry.

However, our purpose here was not demonstration of effectiveness of these definitions or effectiveness of QSAR. The interested reader can learn it from papers [3, 4] and references inside of these papers. As we said above, we use QSAR only for clearness; the proof is correct for any field of use of Naive Bayes classifier.

Author details

Oleg Kupervasser

Address all correspondence to: olegkup@yahoo.com

Department of Mathematics, Ariel University, Ariel, Israel

References

[1] Kupervasser O. The mysterious optimality of Naive Bayes: Estimation of the probability in the system of "classifiers". Pattern Recognition and Image Analysis. 2014;**24**(1):1-10. http://arxiv.org/abs/cs/0202020v1; 2002

[2] Maslennikov ED, Sulimov AV, Savkin IA, Evdokimova MA, Zateyshchikov DA, Nosikov VV, et al. An intuitive risk factors search algorithm: Usage of the Bayesian network technique in personalized medicine. Journal of Applied Statistics. 2015;**42**(1):71-87

[3] Ramensky V, Sobol A, Zaitseva N, Rubinov A, Zosimov V. A novel approach to local similarity of protein binding sites substantially improves computational drug design results. Proteins: Structure, Function, and Bioinformatics. 2007;**69**(2):349-357

[4] Nikitin S, Zaitseva N, Demina O, Solovieva V, Mazin E, Mikhalev S, et al. A very large diversity space of synthetically accessible compounds for use with drug design programs. Journal of Computer-Aided Molecular Design. 2005;**19**:47-63

[5] Raymer ML, Doom TE, Kuhn LA, Punch WF. Knowledge discovery in medical and biological datasets using a hybrid Bayes classifier/evolutionary algorithm. IEEE Transactions on Systems, Man, and Cybernetics. 2003;**33B**:802

[6] Domingos P, Pazzani M. On the optimality of the simple Bayesian classifier under zero-one loss. Machine Learning. 1997;**29**:103

[7] Zhang H. The optimality of Naive Bayes. In: FLAIRS Conference; 2004. Available from: http://www.cs.unb.ca/profs/hzhang/publications/FLAIRS04ZhangH.pdf

[8] Kuncheva LI. On the optimality of Naive Bayes with dependent binary features. Pattern Recognition Letters. 2006;**27**:830

[9] Landau LD, Lifshitz EM. Statistical Physics. Vol. 5. United Kingdom: Elsevier Science Technology; 1996

Chapter 8

Bayesian Graphical Model Application for Monetary Policy and Macroeconomic Performance in Nigeria

David Oluseun Olayungbo

Additional information is available at the end of the chapter

http://dx.doi.org/10.5772/intechopen.87994

Abstract

This study applies Bayesian graphical networks (BGN) using Bayesian graphical vector autoregressive (BGVAR) model with efficient Markov chain Monte Carlo (MCMC) Metropolis-Hastings (M-H) sampling algorithm in a dynamic interaction among monetary policies and macroeconomic performances in Nigeria for the period of 1986Q1–2017Q4. The motivation stems from the instability in the movement of exchange rate, inflation rate and interest rate in Nigeria over the past years as a result of the structure of the economy. In this way, the monetary authority periodically applies the various policy instruments to stabilize the economy using reserve and money supply as at when due. This study adapts VAR and SVAR structure to examine the dynamic interaction among variables of interest, using BN, to provide a better understanding of the monetary policy dynamics and fit the changing structure of the Nigeria's economy as regards the dynamics in her economic structure. Our results show that inflation is the strong predictor of interest rate in Nigeria. A monetary policy of broad inflation targeting is recommended for the country.

Keywords: Bayesian graphical networks, SVAR, MCMC, M-H, Granger-causal inference, Nigeria

1. Introduction

A network can be described as a set of items, with nodes or vertices that are related by edges or links for a specific purpose. There are different types of networks. These networks can be social, economic, informational, technological, biological and so on. However, in this paper, we are interested in Bayesian graphical network (BGN) to investigate causal inferences among monetary policies and macroeconomic performances in Nigeria. Causal effects have been used in economic literature starting from Granger [1], Engle and Granger [2] and Sims [3] using vector autoregression (VAR) to Amisano and Giannini [4] and Blanchard and Quah [5] using structural vector autoregression (SVAR). The progression from VAR to SVAR has been a result of over-parameterization and identification problem associated with VAR that have limited its use for forecasting [6]. SVAR, in a way, has been able to overcome these problems through the application of the recursive and the non-recursive structural model [5]. The advantage of the BGN is that there are directed acyclic graphs (DAG) with nodes or vertices called variables having edges that indicate structural dependence among the variables of interest. Studies that have used graphical models to causal relationships are Pearl [7], Spirites et al. [8], Demiralp and Hoover [9] and Moneta [10], among others. Our intention is to combine BGN with SVAR, in order to examine interrelationship among monetary policies and macroeconomic performances in Nigeria. However, many previous studies have worked on causal inference using BGN-SVAR in developed countries, studies like Swanson and Granger [11], Demiralp and Hoover [9], Moneta and Spirites [12], Corander and Villani [13] and more recently Ahelegbey et al. [6]. The advantage of the BGN-SVAR is that causal influences are tracked down among variables of interest either instantaneously or with time lags using conditional probabilistic inference on the structural model.

Importantly, in this present study, we adopted the BGN-SVAR approach used by Ahelegbey et al. [6] and deal with the identification structure to derive the Bayesian networks using DAG. The Bayesian structural model is then simulated using the multi-move Markov chain Monte Carlo Metropolis-Hastings sampling algorithm. The analyses are done with a view to examine the causal relationship among monetary policy actions and macroeconomic performances in Nigeria. The BGN-SVAR method is more superior to the usual standard Granger causality test, thereby providing an important tool for policy implications, especially for an emerging country like Nigeria where monetary policy stance is a major factor to the macroeconomic performance of the economy. The paper is as follows. Section 2 gives the data source, variable definition and the descriptive statistics, and Section 3 provides an overview of the BGN-SVAR model. While Section 4 highlights the empirical discussion, Section 5 concludes and states the policy recommendation.

2. Data source and variable definition

All the data used for this study were sourced from the Statistical Bulletin [14] published by the Central Bank of Nigeria. The data range from the first quarter of 1986 to the fourth quarter of 2017. The choice of the scope of study is strictly informed by data availability. Monetary policy variable is measured using broad money supply normally referred to as M2, which is defined as the sum of currency in circulation, demand deposit and time deposit in the banking sector. The industrial output (Ind) comprises of crude oil petroleum, natural gas, solid minerals, coal mining, metal ores, quarrying, mining, manufacturing, oil refining and cement production. In addition, interest rate is measured with lending rate (Intr) on banks' credit to the public. Inflation rate (Infl) is measured by the average price index of consumer goods over the period of study. Exchange rate (Exch), on the other hand, is the rate of change of the local currency, naira, to the United State (US) dollar. The industrial output and money supply are measured in the local currency, naira, while the exchange rate is measured in US dollars.

2.1. Descriptive statistics of variables

The descriptive statistics as presented in **Table 1** shows industrial output, exchange rate, inflation rate, interest rate and money supply variables in their unit form with 124 observations for the period of study. The difference between the average mean value and the maximum value of exchange rate, inflation rate and interest shows the high volatility of these variables. The average values for exchange rate, inflation rate and interest rate are 93.86, 19.25 and 14.1%, respectively, while the maximum values are 304.72, 73.1 and 26.7%. The differences between the average values and the minimum values also support the volatility behaviour of the variables. In addition, the volatility movement in these variables as shown in **Figure 1** equally confirms their fluctuation. The skewness of all the variables implies positive skewness of the data distribution. Finally, the significance of the probability of the Jarque-Bera at 5% indicates that all the variables reject the acceptance of the null hypothesis of normal distribution except for the exchange rate variable.

2.2. Unit root test

In order to ascertain the order of integration of our variables, we used both augmented [15, 16] tests. The results from **Table 2** show that all the variables are nonstationary using ADF test. The PP, on the other hand, shows all the variables to be nonstationary except industrial output. Given the stationarity of industrial output with PP, we further conducted structural break following Perron [17] in that its presence can bias the unit root result. The structural break results show that all the variables are nonstationary and support the ADF test. The structural

	Industrial output	Exchange rate	Inflation rate	Interest rate	Money supply
Mean	32,497.84	93.86	19.25	14.1	4,916,951
Median	30,995.49	117.16	11.25	13.66	1,294,950
Maximum	52,931.79	304.72	73.1	26.7	23,388,300
Minimum	18,998.23	3.76	2.14	6	22,730.8
Std. dev	7778.25	69.49	18.1	4	6,665,696
Skewness	0.53	0.24	1.5	0.39	1.244417
Kurtosis	2.65	2.53	3.88	3.87	3.17
Jarque-Bera	6.53	2.36	50.67	7.17	32.15
Prob.	0.03	0.31	0.00	0.03	0.00
Sum	4030	11,638.86	2386.88	1748.6	6.10E + 08
Sum sq. dev	7.44E + 09	593,982.6	4.03E+04	1977.4	5.47E + 15
Observation	124	124	124	124	124

Table 1. Descriptive statistics for selected variables.

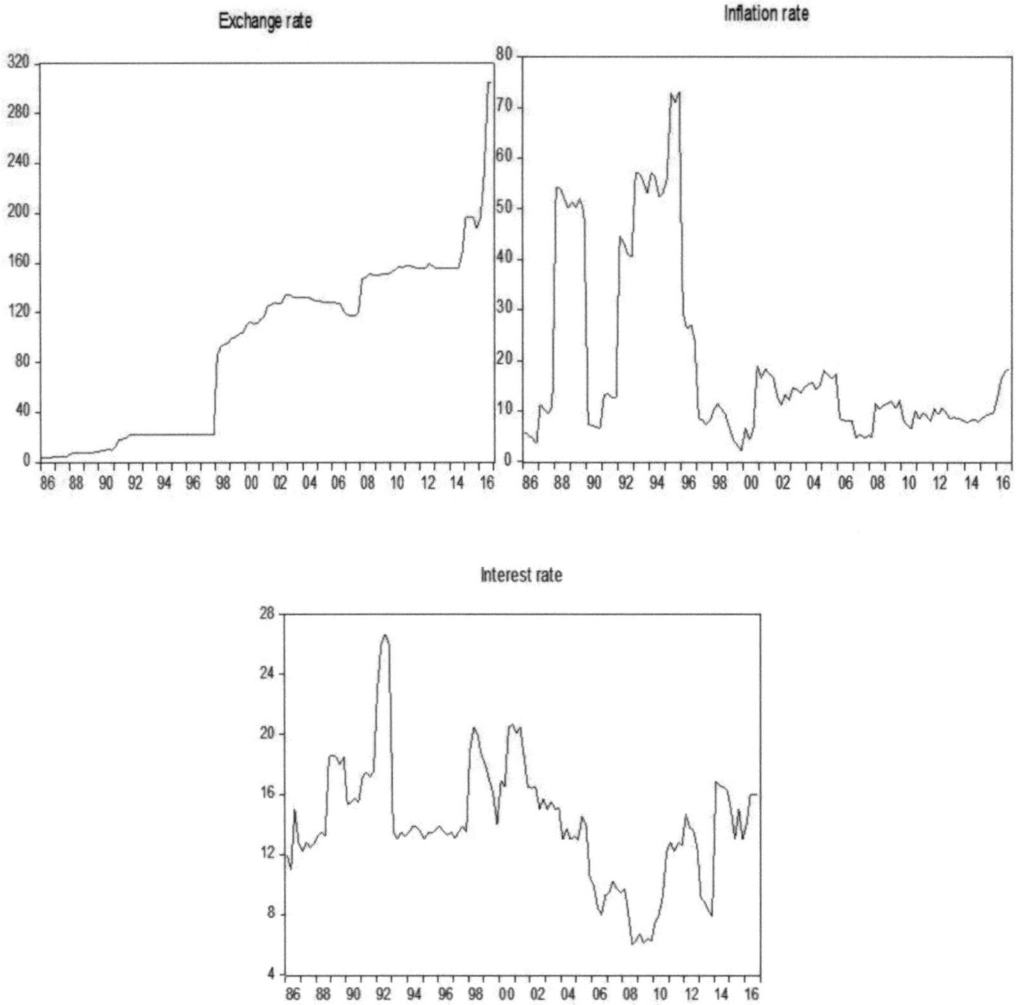

Figure 1. The volatility behaviour of exchange rate, inflation rate and interest rate.

break result is not presented here but available upon request. Given the nonstationarity of all the variables, we proceeded to examine their cointegrating relationship over the period of study.

2.3. Cointegration test

We followed Johansen [18] cointegration technique that compares the trace and the eigenvalue with their critical values for the rejection or acceptance of the null hypothesis of no cointegration. The optimal lag lenght of 1 was chosen following Schwarz Criterion's (SC) result in **Table A4** at the Appendix. The cointegration results presented in **Table 3** show that the trace

	Augmented Dickey-Fuller		Phillips-Peron		
Variables	**Levels**	**First diff.**	**Variables**	**Level**	**First diff.**
Ind	−1.7046	−6.2244	Ind	−4.2882	—
Exch	0.8368	−8.4777	Exch	1.2655	−8.3739
Infl	1.4409	−6.4756	Infl	−2.8021	−10.695
Intr	−2.8834	−10.4781	Intr	−2.7495	−11.2467
M2	5.1489	−11.3326	Mss	5.5596	−11.7113

Note: The critical values at 1, 5 and 10%, respectively, are −3.4846, −2.8853 and −2.5749 for both ADF and PP. Ind, Exch, Infl, Intr and M2 indicate industry output, exchange rate, inflation rate, interest rate and money supply, respectively.

Table 2. Unit root tests.

Coint. rank	Eigenvalue	Trace stat.	Critical value	Prob.
$r = 0$	0.44	108.64	69.82	0.00
$r \leq 1$	0.13	37.29	47.86	0.33
$r \leq 2$	0.09	19.89	29.8	0.43
$r \leq 3$	0.05	8.28	15.5	0.44
$r \leq 4$	0.01	1.79	3.84	0.18
Coint. rank	Eigenvalue	Eigen stat.	Critical value	Prob.
$r = 0$	0.44	71.35	33.88	0.00
$r \leq 1$	0.13	17.39	27.58	0.54
$r \leq 2$	0.09	11.61	21.13	0.58
$r \leq 3$	0.05	6.48	14.26	0.55
$r \leq 4$	0.01	1.8	3.84	0.18

Note: The null hypothesis, H0, of no cointegration is rejected when the value of the trace and maximal eigen statistics is greater than the critical values at 5% significance level.

Table 3. Johansen cointegration results.

statistics is greater than the critical value at 5% significance level at $r = 0$. This hypothesis testing takes us to the next cointegrating vector, $r \leq 1$, where the trace statistics is less than the critical value. We therefore conclude that there is a long-run relationship among the variables.

3. Bayesian graphical VAR model

The starting point of the BGVAR is from VAR process proposed by Sim [3] as dynamic endogenous variables specified as

$$Y_t = B_0 + B_1 Y_{t-1} + B_2 Y_{t-2} + \cdots + B_\rho Y_{t-\rho} + \varepsilon_t \tag{1}$$

Eq. (1) is a vector autoregressive process of order ρ, and it can be respecified in a reduced form as

$$Y_t = B_0 + \sum_{i=1}^{\rho} B_i Y_{t-i} + \varepsilon_t \tag{2}$$

Eq. (2) can further be specified in SVAR process following Amisano and Giannini [4] and Blanchard and Quah [5] with a lower triangular matrix A as

$$AY_t = B_0 + \sum_{i=1}^{\rho} B_i Y_{t-i} + \varepsilon_t \tag{3}$$

Eq. (3) can be written in an inverted form as

$$Y_t = \beta_0 + \sum_{i=1}^{\rho} \beta_i Y_{t-i} + u_t \tag{4}$$

where $\beta_0 = A^{-1}B_0$, $\beta_i = A^{-1}B_i$ and $1 \le i \le \rho$ are lags of the parameter matrices and $u_t = A^{-1}\Sigma\varepsilon_t$ where Σ is a diagonal matrix of variance and covariance matrix. Assume in econometric term that $Y_{t-i} = (X_{t-i})^{'}$; then Eq.(4) can be expressed as

$$Y_t = \beta_0 + \sum_{i=1}^{\rho} \beta_i X_{t-i} + u_t \tag{5}$$

Eq. (5) can be written in matrix form as

$$Y = \beta^{'} X + U \tag{6}$$

The solution to the SVAR model can be achieved through the parameter identification by placing restrictions on the lower triangular matrix A or the B diagonal matrix following the relevant economic theories for the underlying variables. The general method of solving the structural dynamics of the SVAR model after placing the necessary restrictions to attain identification is to determine the effects of the shocks on the contemporaneous variables through the impulse response functions. The impulse response function can be represented through the diagonal matrix of the covariance matrix as

$$\Sigma = A^{-1}\Sigma\varepsilon_t\left(A^{-1}\right) \tag{7}$$

where $\Sigma\varepsilon_t = \varepsilon_t\varepsilon_t^{'}$ and the covariance matrix are assumed to be an identity matrix. The Cholesky decomposition where all the elements above the diagonal are zero (lower triangular matrix) is the usual way of solving the identification problem in SVAR to identify the structural shocks. The system thus becomes exactly identified by comparing the known elements, $\frac{n^2+n}{2}$, and the unknown elements, $n^2 - n + n$, of the covariance matrix [5]. Interestingly, the graphical model can be represented in the form of SVAR following Ahelegbey et al. [6] exposition. A graphical model can be described as the representation of the conditional relationships among random variables. Graphical models are generated in nodes and edges. The nodes house the variables, while the edges point to their relationships. A bivariate graphical model can be written as $X \to Y$, meaning Xcauses Y, where Y (child) is the dependent variable and X (parent) is the independent variable. In a multivariate setting, the graphical model can be expressed as $X \to Y \to Z$ and can be interpreted to mean that the relationship between $X$and Z is conditional on variable Y. Assume $Y_t = \left(Y^1_{t-1}, Y^2_{t-2}, \cdots, Y^n_{t-n}\right)$ where Y^i_t is a realization of the ith variable at time t. The graphical network of a DAG can be written in the form of Eq. (1) as $Y^J_{t-s} \to Y^i_t$ where $B^*_{\rho, ij} \neq 0$ and $0 \leq s \leq \rho$. This implies that past value of Y^J_{t-s} at time lag s, with $(s < t)$, causes the future value of Y^i_t. This explains the notion that cause precedes effect in time. Therefore, following past studies such as Corander and Villani [13] and Ahelegbey et al. [6] among others, a network structure of a DAG is described as $G = (V, A)$, where $V$is a finite set of nodes symbolizing $Y_t = \left(Y^1_{t-1}, Y^2_{t-2}, \cdots, Y^n_{t-n}\right)$ in this case, while A is also a finite set of directed edges denoting $Y^J_{t-s} \to Y^i_t$ stated earlier. In other words, the graphical model can be specified in VAR representation as $B_s = \left(G_S, \varphi_s\right)$, where s is a period lag, B is the structural parameters of the interdependent variable Y_t, G_s is the binary connectivity matrix and φ_s is a matrix of coefficient of lag s. At period s for $0 \leq s \leq \rho$, $G_{s, ij} = 1$ implies a causal effect of $Y^J_{t-s} \to Y^i_t$, and $G_{s, ij} = 0$ implies no causal relationship between Y^J_{t-s} and Y^i_t with $\varphi_{s, ij}$ indicating the quantity of the causal effects of $Y^J_{t-s} \to Y^i_t$. The graphical model can further be written in a VAR form from Eq. (2) as.

$$Y_t = \left(G_0, \varphi_0\right) + \sum_{i=1}^{\rho} \left(G_i, \varphi_i\right) Y_{t-i} + \varepsilon_t \tag{8}$$

Following Koop [19], Geweke [20] and Olayungbo [21], the posterior distribution in Bayes' theorem can be written in continuous form as

$$P(\theta|Y) = \frac{P(Y|\theta)P(\theta)}{\int P(Y|\theta)P(\theta)d\theta} \tag{9}$$

where $P(Y|\theta)$ is the likelihood function and $P(\theta)$ is the prior. In proportionality form, Eq. (9) becomes

$$P(\theta|Y) \propto P(Y|\theta)P(\theta) \tag{10}$$

Given the parameters to be estimated in our models, the posterior becomes

$$P\left(\mu, \Sigma_y^{-1}, G|Y\right) \propto P\left(Y|\mu, \Sigma_y^{-1}, G\right) P\left(\Sigma_y^{-1}|G\right) \tag{11}$$

where $\mu = 0$ and the likelihood function is generated with respect to Eq. (8) as

$$P\left(Y|\Sigma_y^{-1}, G\right) = (2\pi)^{\frac{-nT}{2}}\left|\Sigma_y^{-1}\right|^{\frac{T}{2}}\exp\left\{-\frac{1}{2}\left(\Sigma_y^{-1}, \sum_{t=1}^{T} Y_t Y_t^{'}\right\}\right. \tag{12}$$

The prior density, $P(G)$, is chosen from a uniform prior as $P(G) \propto 1$; the inverse of the variance-covariance matrix of the error term follows a Wishart distribution, i.e., $\Sigma_y^{-1} \sim W\left(\nu, \sum_{t=1}^{T}\left(Y_0 Y_0^{'}\right)^{-1}\right.$; and β is set to 0.5. The prior density is then written as

$$P\left(\Sigma_y^{-1}|G\right) = \frac{1}{K_n(\nu, (Y_0 Y_0)^{'}}\left|\Sigma_y^{-1}\right|^{\frac{(\nu-n-1)}{2}}\exp\left\{-\frac{1}{2}\left(\Sigma_y^{-1}, \sum_{t=1}^{T} Y_0 Y_0^{'}\right\}\right. \tag{13}$$

The posterior distribution is written with the likelihood function and prior density in Eqs. (12) and (13) as.

$$\begin{aligned} P\left(\mu, \Sigma_y^{-1}|Y\right) &\propto (2\pi)^{\frac{-nT}{2}}\left|\Sigma_y^{-1}\right|^{\frac{T}{2}}\exp\left\{-\frac{1}{2}\left(\Sigma_y^{-1}, \sum_{t=1}^{T} Y_t Y_t^{'}\right\}\right. \\ &\cdot \frac{1}{K_n(\nu, (Y_0 Y_0)^{'}}\left|\Sigma_y^{-1}\right|^{\frac{(\nu-n-1)}{2}}\exp\left\{-\frac{1}{2}\left(\Sigma_y^{-1}, \sum_{t=1}^{T} Y_0 Y_0^{'}\right\}\right. \end{aligned} \tag{14}$$

Eq. (14) which is the posterior distribution with the likelihood function and the prior density in Eqs. (12) and (13) is estimated with Markov chain Monte Carlo sampling methods and Metropolis-Hastings recursively to obtain the posterior means. Samples are drawn from the posterior distribution, $P\left(\Sigma_y^{-1}, G|Y\right)$, given $P\left(Y|\Sigma_y^{-1}, G\right)$ and $P\left(\Sigma_y^{-1}|G\right)$ by using the MCMC and M-H algorithm (see Ahelegbey et al. [6] for more exposition).

3.1. Granger causality test

Following Granger [1], a pairwise Granger causal (P-GC) relationship that conditions a variable on other variables and their time lags is investigated. The reality about the P-GC causality in the graphical network analysis is that it is a directed and forward causal relationship among the dependence variable structures. The P-GC VAR (ρ) model is stated as

$$Y_t^i = \sum_{s=1}^{\rho} \alpha_s Y_{t-s}^i + \sum_{s=0}^{\rho} \beta_s Y_{t-s}^j \tag{15}$$

4. Empirical analysis and discussion

Following Ahelegbey et al. [6], we sampled and derive separately the multivariate autoregressive (MAR) and the multivariate instantaneous (MIN) system using M-H algorithm specified in Appendix 2. The MAR network is a contemporaneous interaction of the variables from their past realizations to current realizations, while the MIN network is a contemporaneous

http://dx.doi.org/10.5772/intechopen.87994

interaction of the variables in their current realizations. At the implementation level, we ordered our variables in the Y_t vector as Y_t =(M2, Ind, Infl, Intr, Exch) from Eq. (2). It should be noted that our results are not sensitive to the variable ordering, in which case, any order can be taken.

After choosing an optimal lag length of 1 using Akaike information criterion (AIC) and Schwarz criterion (SC) (see Appendix 1), the BGN-VAR model ran 20,000 Gibbs iteration each for the MAR and MIN, making a total number of iteration to be 40,000, out of which 20,000 was set as burn-ins to achieve convergence. The results of the P-GC VAR for the MIN and MAR are presented in **Figures 2** and **3**. The dark green (light) color implies strong (weak) dependence between the dependent and independent variables. The row variables are the independent or explanatory variables, while the column variables are the dependent or response variables. The result of the MIN in **Figure 2** shows strong evidence of causal relationship with M2→Infl, Ind→Infl, Infl→Intr. The results can be interpreted to mean a stronger evidence of effects of both money supply and industrial output on inflation rate in Nigeria. This implies that the increase of money in circulation increases industrial demand and production; however, the resultant effects lead to inflationary rate due to high cost of production in Nigeria. High cost of production for firms and companies generates increase in prices of goods and services due to the use of generators to produce. The country generates 5000 megawatts (MW) of electricity which is not enough in residential needs let alone industrial needs; hence, most firms result to the use of generating set which results to high cost of production and high prices. We also found stronger effects of inflation rate behaviour on interest rate movement in Nigeria. This implies that instantaneous change in money supply and industrial output are major determinants of inflation rate in Nigeria. We also found strong effects of inflation rate on interest rate. This implies that resultant inflationary effects compel monetary authority and commercial banks to choose high interest rate to stabilize and ensure reasonable returns on investment. Furthermore, **Figure 3** gives the contemporaneous autoregressive structure of the variable of interest. The MAR result shows the causal edges as, firstly, $M2_{t-1}$ →M2, Ind and

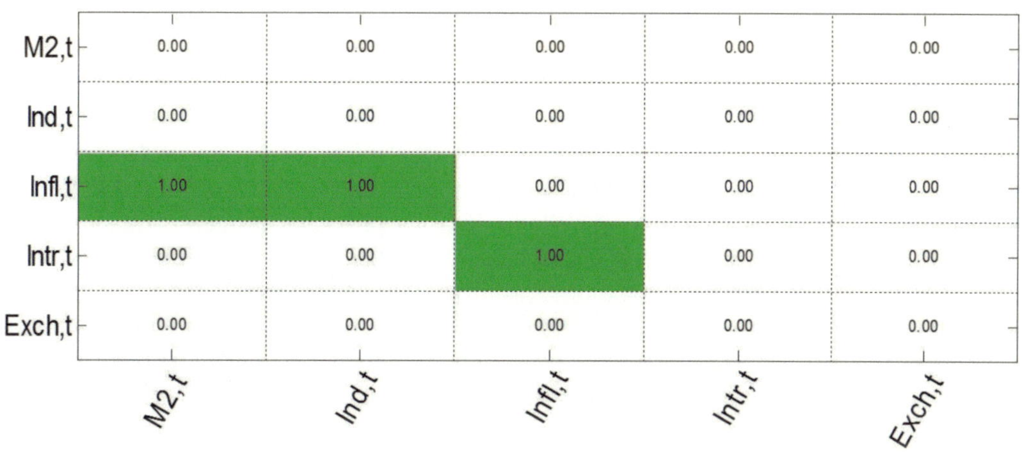

Figure 2. Multivariate instantaneous structure (MIN).

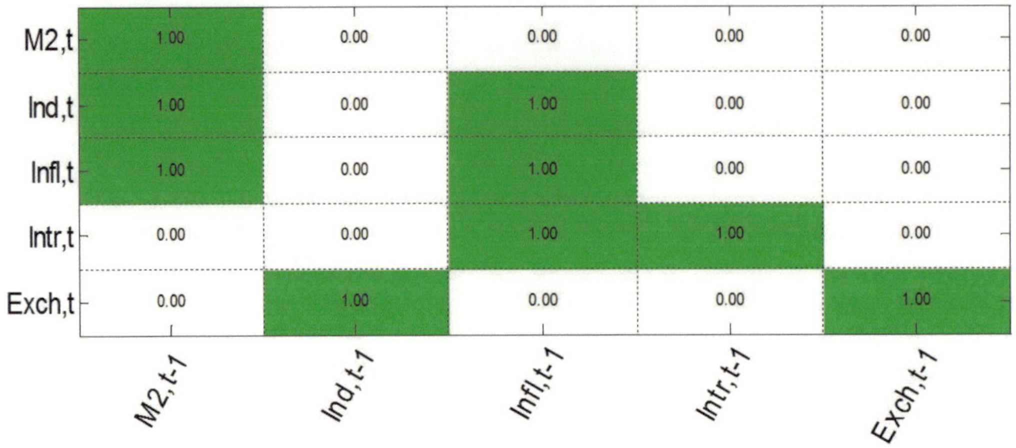

Figure 3. Multivariate autoregressive structure (MAR).

Infl; secondly, Ind_{t-1} →Exch; thirdly, $Infl_{t-1}$ →Intr, Infl and Ind; fourthly, $Intr_{t-1}$ →Intr; and, lastly, $Exch_{t-1}$ →Exch. The causal edges can be interpreted to mean that inflation rate, industrial output and money supply respond strongly to immediate past lag of money supply. Furthermore, we found past value of industrial output to Granger-cause exchange rate. In addition, past value of inflation Granger causes interest rate, current inflation and industrial output. This outcome corroborates the MIN result that inflation rate is a strong predictor of interest rate in Nigeria. The fourth and last causal edges can be explained to mean both past interest rate and exchange rate are strong predictors of their current states.

5. Conclusion and policy recommendations

This study examines the dynamic interactions among monetary policies and macroeconomic performances in Nigeria over the period of 1986Q1–2017Q4 with the application of BNG-SVAR. The use of the BNG-SVAR comes from the dynamic response of monetary policy to macroeconomic indicators in Nigeria. The nonstationary process of the data used led to test for their cointegration. The cointegration results show the existence of long-run relationship among the variables of interest. The P-GC results from the BNG-SVAR with the MCMC and M-H sampling techniques show that inflation is a strong predictor of interest rate in Nigeria given both the contemporaneous instantaneous (MIN) and the contemporaneous autoregressive (MAR) results. This is more reason why the Central Bank of Nigeria (CBN) in its period monetary policy committee (MPC) targets the inflation rate by choosing appropriate monetary policy rate (MPR) in response to the inflation rate in the country. This study, therefore, recommends that the inflation targeting in Nigeria should be broad and not limited to changing the MPR only. The fiscal discipline should be ensured from the Ministry of finance and the executive arm of the government. In any case, any fiscal policy should be directed towards the productive sectors of the economy. A major determinant of inflationary pressure in Nigeria is incessant supply of electricity. There should be massive investment in power generation and transmission in the country to eliminate the additional cost of production that drives prices up through the use of generating sets for production. Finally, the exchange rate policy should be to ensure domestic production to drive prices down rather than reliance on imported goods that promote imported inflation.

A. Appendices

A.1. Appendix 1

See the **Table A4**.

VAR lag order selection criteria					
Endogenous variables: M2, Ind, Infl, Intr, Exch					
Lag	**LogL**	**LR**	**FPE**	**AIC**	**SC**
0	−4973.01	NA	7.41e + 29	82.96683	83.08297
1	−4220.51	1429.753	4.02e + 24*	70.84180*	71.53868*
2	−4206.64	25.18928	4.85e + 24	71.02738	72.30498
3	−4190.38	28.18071	5.64e + 24	71.17307	73.03140
4	−4168.07	36.81222	5.96e + 24	71.21790	73.65696
5	−4122.25	71.78438*	4.28e + 24	70.87090	73.89069
6	−4100.89	31.68685	4.66e + 24	70.93154	74.53205
7	−4081.64	26.94965	5.32e + 24	71.02738	75.20861
8	−4056.97	32.48824	5.61e + 24	71.03280	75.79476

*Indicates the optimal lag length, where LogL, LR, FPE, AIC and SC indicate log likelihood, likelihood ratio, final prediction error, Akaike information criterion and Schwarz criterion.

Table A4. Optimal lag selection results

A.2. Appendix 2

The MCMC and the M-H algorithms are a proposal distribution to sample a new graph G^*conditioned on a graph G with acceptance probability given as.

$M(G^*|G) = \min\left\{\frac{P(Y|G^*)}{P(Y|G)}\frac{P(G^*)}{P(G)}\frac{Z(G|G^*)}{Z(G^*|G)}, 1\right\}$ where $P(Y|G)$ is the likelihood function, $P(G)$ is the prior density and, finally, $Z(G^*|G)$is the proposal distribution.

A.3. Appendix 3

Inverse Wishart Prior Posterior MCMC

The procedure for Gibbs sampling for the independent-normal Wishart prior is as follows:

1. Draw $G_i^{(k)}$ from the normal $p(G_i|Y, \Sigma)$.
2. Draw $\Sigma_i^{-1(k)}$ from the Wishart $p\left(\Sigma_i^{-1}|Y, G_i\right)$.
3. Repeat steps 1 and 2N (20,000) times, and discard the first N_{burn} (10,000) iterations as burn-ins.

A.4. Appendix 4

Identification structure of the BGN-VAR results.

$$B_{-\rho} = \begin{Bmatrix} -0.9 & 0 & 0 & 0 & 0 \\ 0.6 & 0 & 0.5 & 0 & 0 \\ 0.7 & 0 & -0.5 & 0 & 0 \\ 0 & 0 & 0.5 & 0.7 & 0 \\ 0 & 0.6 & 0 & 0 & 0.6 \end{Bmatrix} B\_0 = \begin{Bmatrix} 0 & 0 & -0.8 & 0 & 0 \\ 0 & 0 & 0 & 0 & 0 \\ 0 & 0 & 0 & 0 & 0 \\ 0 & 0.5 & 0 & 0 & -0.5 \\ 0 & 0 & 0 & 0 & 0 \end{Bmatrix}$$

Author details

David Oluseun Olayungbo

Address all correspondence to: doolayungbo@oauife.edu.ng; doolayungbo@gmail.com

Department of Economics, Obafemi Awolowo University, Ile-Ife, Nigeria

References

[1] Granger CWJ. Investigating causal relations by econometrics models and cross-spectral methods. Econometrica. 1969;**37**:424-438

[2] Engle RF, Granger CWJ. Cointegration and error correction: Representation, estimation and testing. Econometrica. 1987;**55**:251-276

[3] Sims CA. Macroeconomics and reality. Econometrica. 1980;**48**:1-48

[4] Amisano G, Giannini C. Topics in Structural VAR Econometrics. 2nd ed. Berlin: Springer; 1997

[5] Blanchard O, Quah D. The dynamic effects of aggregate demand and supply disturbances. American Economic Review. 1989;**79**:655-673

[6] Ahelegbey DF, Billio M, Casarin R. Bayesian graphical models for structural vector auto regressive processes. Journal of Applied Econometrics. 2016;**31**:357-386

[7] Pearl J. Causality: Models, Reasoning and Inference. London, UK: Cambridge University Press; 2000

[8] Spirites P, Glymour C, Scheines R. Causation, Prediction, and Search. Cambridge, MA: MIT Press; 2000

[9] Demiralp S, Hoover KD. Searching for the causal structure of a vector autoregression. Oxford Bulletin of Economics and Statistics. 2003;**65**:745-767

[10] Moneta A. Graphical causal models and VARs: An empirical assessment of the real business cycles hypothesis. Empirical Economics. 2008;**35**(2):275-300

[11] Swanson NR, Granger CWJ. Impulse response functions based on causal approach to residual orthogonalization in vector autoregression. Journal of the American Statistical Association. 1997;**92**:357-367

[12] Moneta A, Spirtes P. Graphical models for the identification of causal structure in multivariate time series models, joint conference on information sciences proceedings. Atlantis press; 2006

[13] Corander J, Villian M. A Bayesian approach to modelling graphical vector autoregression. Journal of Time Series Analysis. 2006;**27**(1):141-156

[14] Statistical Bulletin, Abuja, Nigeria: Central Bank of Nigeria; 2018

[15] Dickey DA, Fuller WA. Distribution of the estimators for autoregressive time series with a unit root. Journal of the American Statistical Association. 1979;**74**:427-431

[16] Phillips PCB, Perron P. Testing for a unit in time series regression. Biometrica. 1988;75: 335-346. DOI: org/10.1093/biomet/7.52.335.

[17] Perron P. The great crash, the oil price shock and the unit root hypothesis. Econometrica. 1989;**57**:1361-1401

[18] Johansen S. Statistical analysis of cointegration vectors. Journal of Economic Dynamics and Control. 1988;**12**:231-254

[19] Koop G. Bayesian Econometrics. Hemel Hempstead: Wiley-Interscience; 2003

[20] Geweke J. Contemporary Bayesian Econometrics and Statistics. New York: Wiley; 2005

[21] Olayungbo DO, Akinlo AE. Insurance penetration and economic growth in Africa: Dynamic effects analysis using Bayesian TVP VAR approach. Cogent Economics and Finance. 2016;**4**:1150390. http://dx.doi.org/10.1080/23322039.2016.1150390